SpringerBriefs in Bioengineering

SpringerBriefs present concise summaries of cutting-edge research and practical applications across a wide spectrum of fields. Featuring compact volumes of 50 to 125 pages, the series covers a range of content from professional to academic. Typical topics might include: A timely report of state-of-the art analytical techniques, a bridge between new research results, as published in journal articles, and a contextual literature review, a snapshot of a hot or emerging topic, an in-depth case study, a presentation of core concepts that students must understand in order to make independent contributions.

More information about this series at https://link.springer.com/bookseries/10280

Manfred Raff

Mass Transfer Models in Membrane Processes

Applications in Artificial Organs

 Springer

Manfred Raff
Bisingen, Baden-Württemberg, Germany

ISSN 2193-097X   ISSN 2193-0988   (electronic)
SpringerBriefs in Bioengineering
ISBN 978-3-030-89194-7   ISBN 978-3-030-89195-4   (eBook)
https://doi.org/10.1007/978-3-030-89195-4

Preface

The scientific focus of my professional life is membrane technology. It started 1976 at University of Stuttgart, Germany, with a cooperative project between university, nephrology and industry, during which I developed a prototype of a new dialysis machine, which should be steam sterilisable, to be used with a car battery and avoid leakages in piston pumps.

As an employee of Gambro Dialyzers, Germany (since 2012 Baxter), between 1982 and 1991, I was involved in research, development and production of membranes and modules for kidney and liver dialysis, plasmapheresis and oxygenation.

Between 1991 and 2014, I worked as Professor at Furtwangen University, first in the "Department of Process Engineering", and later as Dean of the faculties "Mechanical and Process Engineering" and "Medical and Life Sciences". In cooperation with industry, during that time I carried out various membrane projects and offered the topic of membrane processes in elective and compulsory lectures in the bachelor and master education. Since 2014, I am retired but still working for 2 further years as Assistant Professor at the Cooperative State University Stuttgart (lecture: "process engineering") and for further 4 years at Furtwangen University (lecture: "artificial organs").

I would like to express my sincere thanks to the companies Baxter and Fresenius for their support with literature, brochures and pictures for this book. Many thanks also to all scientists and publishers, supporting me by the permission to reprint figures from their publications.

And last but not least, I would especially thank my wife Monika for always actively supporting me in living my professional dreams.

Bisingen, Germany   Manfred Raff

Contents

Symbols

A^m	Membrane surface area (m^2)
A^s	Surface of polymer particles (m^2)
A^g	Cross section of membrane pore (m^2)
A_q	Cross section of flow channel (m^2)
c	Concentration (g/l)
Δc_m	Mean logarithmic concentration difference (g/l)
Cl	Clearance (ml/min)
d_i	Inner diameter of a hollow fibre membrane (m)
d_h	Hydraulic diameter (m)
d_E	Equivalent diameter (m)
D	Diffusion coefficient (m^2/s)
E	Enhancement factor
f	Free fraction of a component in the plasma
He	Henry coefficient (Pa)
k	Boltzmann constant ($= 1.381 \cdot 10^{23}$ kg $\cdot$ m^2/(K $\cdot$ s^2))
K_0	Overall mass transfer coefficient (m/s)
K_A	Equilibrium constant (M^{-1})
L_{eff}	Length of the hollow fibre membranes between the potting (m)
L_p	Hydraulic permeability (m/(s $\cdot$ Pa))
$\check{M}$	Molar mass (g/mol)
$\dot{m}$	Mass flow density (kg/s/m^2)
$\dot{M}$	Mass flow (kg/s)
N	Number of hollow fibre membranes in the module
p	Pressure (Pa)
P_i^m	Diffusive permeability of comp. i in the membrane (m/s)
Q	Volume flow (ml/min)
r_i	Molecular radius of comp. i (m)
R_i	Rejection coefficient of comp. i
R^b	Mass transfer resistance of blood boundary layer (s/m)
R^d	Mass transfer resistance of dialysate boundary layer (s/m)

R^m	Mass transfer resistance of membrane phase (s/m)
R_h^m	Hydraulic membrane resistance (1/m)
$\tilde{R}$	Universal gas constant ($= 8.314$ J/(mol · K))
S_i	Sieving coefficient of comp. i
S	Hb-O_2 saturation
TMP	Transmembrane pressure (mmHg), (Pa)
U	Circumference (m)
UF	Ultrafiltration (ml/min)
UFC	Ultrafiltration coefficient (ml/(min · mmHg))
$\dot{v}$	Volume flow density (flux) (m/s)
$\dot{V}$	Volume flow (m³/s)
w	Flow velocity (m/s)
x	Coordinate parallel to the membrane surface
$\tilde{x}_i$	Molar fraction of comp. i in the liquid phase
$\tilde{y}_i$	Molar fraction of comp. i in the gas phase
z	Coordinate perpendicular to the membrane surface
Δz^m	Thickness of the membrane (m)
Z_{por}^m	Effective length of the pores in the membrane (m)

Dimensionless Parameters

Re	Reynolds number
Sc	Schmidtzahl
Sh	Sherwood number

Greek Letters

a	Bunsen coefficient (ml/(l · mmHg))
β	Mass transfer coefficient (m/s)
δ	Boundary layer thickness (m)
μ^m	Membrane—tortuosity
ε^m	Membrane—porosity
λ	Friction factor
ρ	Density (kg/m³)
η	Dynamic viscosity (Pa · s)
γ	Shear rate (1/s)
ν	Kinematic viscosity (m²/s)
π	Osmotic pressure (Pa)

Subscripted Indices

A	Albumin
out	At the exit
c	Related to concentration
in	At entry
CO_2	Carbon dioxide
GG	Balance
h	Hydraulic
H_2O	Water
i	Inside/inner
i	Component i (solute)
j	Component j (solvent)
b	Bulk (concentration) or bound (to Protein)
m	Mean value of …
O_2	Oxygen
por	Pores
PR	Phenol red
q	Cross section
x	Membrane element location
w	Wall (concentration)

Superscript Indices

b	Blood
d	Dialysate
f	Feed
g	Gas
l	Liquid
m	Membrane
p	Permeate
s	Solid
v	Volume (-concentration)

Chapter 1
Introduction

Abstract Membrane processes for artificial organs (AO) ensure that the state of equilibrium (homeostasis) in the blood circulation is maintained by exchanging/eliminating blood components via semi-permeable membranes. This requires an extracorporeal circuit that is controlled and monitored with appropriate hardware and in which the blood is pumped through a membrane module where the exchange of substances takes place. Patients are connected to the module via needles and lines, disposables, which are usually used only for one treatment.

According to the statistics of Eurotransplant at the end of 2020 in Germany, 9463 people have been on the waiting list for a donor organ. Most patients, 7338, were waiting for a kidney, 891 for a liver, 700 for a heart and 279 for a lung. With 913 post-mortal organ donors (2941 organs) nationwide, there was an acute shortage which, without the use of artificial organs (AO), would have meant death for many patients with chronic and acute organ failure (see: https://www.dso.de/SiteCollectionDocuments/DSO-Jahresbericht%202020.pdf).

Besides a quite high number of chronically ill patients (at the end of 2020 in Germany about 100.000 dialysis patients), COVID-19 pandemic has dramatically increased the need for AO, as critically ill patients may develop multiple organ dysfunction (MOF) and therefore require extracorporeal multi-organ support (ECOS). As shown in Fig. 1.1, the following intermittent or continuous procedures are used:

- Extracorporeal membrane oxygenation (ECMO) and extracorporeal CO_2 removal ($ECCO_2R$) to support lung function,
- Plasma filtration and exchange (PF/PE) in combination with continuous renal replacement therapies (Continuous Veno-Venous Hemofiltration/Haemodialysis, CVVH/D) to support renal functions and
- Combination of membrane and adsorption processes to support liver functions in the molecular adsorbent recirculation system (MARS) and PROMETHEUS.

Through food intake, respiration and biochemical processes in the blood, concentrations of blood components change and are regulated to normal values by healthy

M. Raff, *Mass Transfer Models in Membrane Processes*,
SpringerBriefs in Bioengineering,
https://doi.org/10.1007/978-3-030-89195-4_1

1

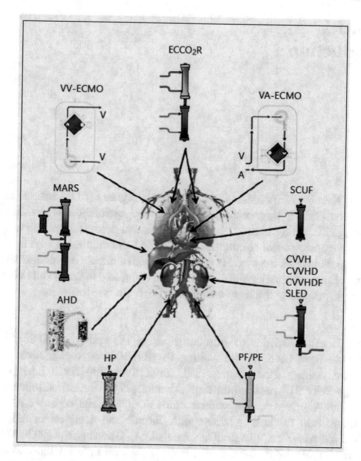

Fig. 1.1 Schematic representation of a different extracorporeal organ support (ECOS) (*Source* Ronco (2021)). Printed with kind permission of Karger

organs, thus ensuring homeostasis, the state of equilibrium, of the dynamic system of blood circulation. For this purpose, it is necessary that an exchange of substances takes place in the organs between the internal blood circulation and organ-specific discharge systems. The selection of components to be exchanged in an organ is ensured by semi-permeable biological membranes, like the basal membranes in the glomeruli of the kidney and in the alveoli of the lung.

Since blood is a disperse mixture (cells in plasma), development for separation processes in AO can draw on experience from technical applications in process engineering. For example, from crossflow microfiltration for the separation of micro-organisms from liquids in biotechnology correlations may be used, as they are needed in plasmapheresis and plasma donation to estimate the dependencies of plasma filtration rate on relevant parameters (see Chaps. 3.3 and 4.3). If dissolved components in blood plasma are to be exchanged, as in the case of lung, kidney and liver, molecular

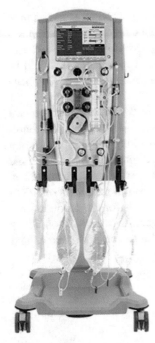 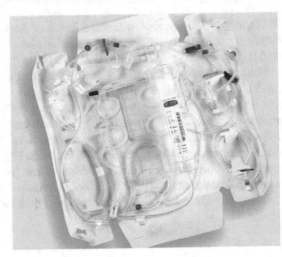

Fig. 1.2 Monitor for "critical care patients" (left photograph: Prismaflex TPE Brochure: USMP/MG 120/18–0021(1) and Prismaflex Hemofilter Set (right photograph: 306100279_1 2009.09. Gambro Lundia AB). Printed with kind permission of Baxter Inc

interactions dominate, which are described with the laws of physical chemistry and thermal process engineering. Thus, for example, the physical solubility of oxygen O_2 and carbon dioxide CO_2 in plasma as well as chemical interaction with haemoglobin may be estimated, which is required to calculate mass transfer rates in an *artificial lung* (see Chap. 4.4).

All extracorporeal processes use monitors, the *hardware*, which are equipped with aggregates (peristaltic pumps, valves, heat exchangers, etc.) with sensors (for temperature, pressure, flow, etc.) and with electronics for MSR tasks to ensure safe circulation of the blood outside the body.

The *software (disposables)* refers to products that are usually only approved for one treatment, such as bloodlines with needles for accessing the patient, and the membrane modules, in which the desired exchange takes place. As an example, Fig. 1.2 shows *hardware* (the monitor) and *disposables* (hemofilter, bloodlines, needles, etc.) from Baxter for the treatment of patients with acute renal failure by Continuous Veno-Venous Hemofiltration (CVVH) (see Chap. 4.1.2).

The functions of monitors are only briefly mentioned in this book in connection with the description of the respective process. The focus here is an introduction to membrane processes and their application to AO, explaining the organ-specific functions and their implementation with membrane modules.

The selection of membranes is based on the separation characteristics required for an organ and on the performance of a membrane module, such as its removal rate for specific components (see, e.g., Ronco (2018), Donato (2017), Boschetti-de-Fierro (2013)). But also, biocompatibility of the membrane material should be further improved to reduce or even avoid inflammatory reactions (see, e.g., Deppisch (2001), Tijink (2013), Irlan (2015), Kohlova (2018)).

Using suitable mass transfer models, it is possible to derive how performance data for an AO depend on the geometric data in the module (membrane area, diameter, length of hollow fibre membranes, etc.), on the operating conditions during a treatment (blood flow, dialysate flow, transmembrane pressure, etc.) and on specific data of the dissolved and dispersed components and the solvent (concentrations, viscosity, haematocrit, diffusion coefficient, etc.). This enables producers of AO to estimate cost/performance ratios and development potentials for the products, and users, medical doctors and patients, to optimise patient-specific treatment conditions via the data obtained from blood analysis and their influence on performance in a process.

References

Boschetti-de-Fierro A, Voigt M, Storr M, Krause B (2013) Extended characterization of new class of membranes for blood purification: the high cut-off membranes Int. J Artif Organs 36(7):455–463

Deppisch RM, Beck W, Goehl H, Ritz E (2001) Complement components as uremic toxins and their potential role as mediators of microinflammation. Kidney Int 50(78):271–277

Donato D, Boschetti-de-Fierro A, Zweigart C, Kolb MEllot S, Storr M, Krause B, Leypoldt K, Segers P (2017) Optimization of dialyzer design to maximize solute removal with two-dimensional transport model. J Membr Sci 519–528

Irfan M, Idris A (2015) Overview of PES biocompatible/hemodialysis membranes: PES-blood interactions and modification techniques. Mater Sci Eng C56:574–582

Kohlova M, Amorim CG, Araujo A, Santos-Silva A, Solich P, Conceicao BS, Montenegro M (2018) The biocompatibility and bioactivity of hemodialysis membranes and their impact in end-stage renal disease. J Artif Organs. https://doi.org/10.1007/s10047-018-1059-9

Ronco C, Bagshaw SM, Bellomo R, Clark WR, Hussein-Syed F, Kellum JA, Ricci Z, Rimmele T, Reis T, Ostermann M (2021) Extracorporeal blood purification and organ support in the critically ill patient during COVID-19 pandemic: expert review and recommendation. Blood Purif 50:17–27

Ronco C, Clark WR (2018) Haemodialysis membranes, Nat Rev. Nephrol 14:394–410

Tijink MSL, Wester M, Glorieux G, Gerritsen KGF, Sun J, Swart PC, Borneman Z, Wesseling M, Vanholder R, Joles JA, Stamatialis D (2013) Biomaterials 34:7819–7828

Chapter 2
Membranes and Modules

Abstract Modules for AO are usually housings equipped with hollow fibre membrane bundles in which blood is adjusted to physiologically necessary (healthy) values using different processes. The requirements for membranes and modules are based on the desired treatment times and on the specific properties of the substances to be exchanged. The description of the characteristic properties of a membrane is therefore based on its rejection characteristics and on the achievable permeate mass flows. To understand how manufacturers may influence the performance and cost-effectiveness of the membranes produced, this chapter briefly presents the phase inversion process to produce hollow fibre membrane bundles and the essential steps for further processing of the bundles into membrane modules.

Membranes, as they are used in AO, nowadays mainly have the form of a tube, a hollow fibre, and are built into housings as bundles with many thousands of fibres. Figure 2.1 shows the schematic of a hollow fibre module in which one hollow fibre represents the entire membrane bundle. By casting the hollow fibres, for example with polyurethane (PUR), against each other and towards the housing, two spaces separated by the membrane are formed in the module housing, the feed/retentate area and the filtrate/permeate area.

The surface area of all hollow fibres facing the blood is selected as the membrane area A^M, which is calculated from the inner surface of the hollow fibres when the blood flows through them:

$$A^M = N \cdot \pi \cdot d_i \cdot L_{\text{eff}} \tag{2.1}$$

N is the number of hollow fibres in the module, d_i the inner diameter and L_{eff} the effective length of the hollow fibres in between the potting areas.

In the application shown in Fig. 2.1, blood to be purified flows on the feed side in the module housing, is distributed into the hollow fibres, which are embedded in the potting material, and flows through them to the retentate side, from which the purified blood returns to the patient. If, as in haemodialysis, a washing fluid (dialysate) flows through the filtrate chamber in counter-current to the blood, a permeate volume flow

© The Author(s), under exclusive license to Springer Nature Switzerland AG 2022
M. Raff, *Mass Transfer Models in Membrane Processes*,
SpringerBriefs in Bioengineering,
https://doi.org/10.1007/978-3-030-89195-4_2

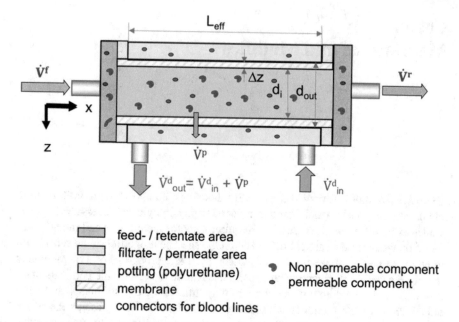

Fig. 2.1 Components, geometry and flow rates in a hollow fibre device during a counter-current process like dialysis

from blood to dialysate will be discharged together with the outgoing dialysate flow ($\dot{V}^d_{out} = \dot{V}^d_{in} + \dot{V}^p$). Consequently, haemodialysis is a *counter-current process* in which dissolved, permeable components are transported due to a concentration gradient across the membrane mainly by diffusion from blood to dialysate. If also plasma water ($\dot{V}^p$) must be eliminated from the patient, there will be an additional small convective transport of dissolved components together with the permeate volume flow (see Chap. 4.1.1).

In hemofiltration (see Chap. 4.1.2), there is no washing fluid on filtrate side, but a much higher permeate flow as compared to dialysis will be necessary, because permeable dissolved components in this process may only be removed by convection. Since permeate is removed in a direction perpendicular to the blood flow, hemofiltration is a *crossflow filtration process*.

For the transport models as described in Chap. 3, the coordinate parallel to the membrane surface is defined as x-direction, whereas the coordinate perpendicular to the membrane surface as z-direction. Consequently, Δz is the thickness of the hollow fibre wall.

2.1 Membrane Properties

A membrane is a thin, semi-permeable layer which connects spaces with each other via porous or non-porous membranes and changes the composition of the adjacent phases through transmembrane transport. For applications in AO, the membranes must be designed in such a way that valuable blood components such as proteins and cells are rejected, and harmful substances such as urea in dialysis and CO_2 in oxygenation are removed from the blood. Some substances such as bicarbonate in dialysis and O_2 in oxygenation are transported from the adjacent phase into the blood.

To describe the semi-permeability of a membrane, one defines the *rejection coefficient* R_i and the *sieving coefficient* S_i of a component i by the following equation:

$$R_i = 1 - c_i^p/c_i^f = 1 - S_i \qquad (2.2)$$

If the permeate concentration c_i^p reaches the feed concentration c_i^f, R_i becomes zero, which means that this component is completely permeable ($S_i = 1$). If a component is completely rejected, the permeate concentration is zero and $R_i = 1$. To describe the membrane characteristics according to this criterion, the rejection of selected molecules and ions of different dimensions are determined experimentally and plotted against their molecular weights. The *molecular weight cut-off* (MWCO) is defined by the molar mass of the component that results in a rejection coefficient of $R_i = 0.9$. The rejection characteristic of a membrane with a MWCO of 60 kD ($1\ D = 1$ Dalton $= 1$ g/mol) is shown in Fig. 2.2. Thus, using such a membrane, more than 90% of molecules with a molar mass higher than 60,000 g/mol will be rejected. Molecules with molar masses smaller than about 100 g/mol pass through this membrane unhindered because they are smaller than the smallest pores. Molecules with molar masses between 100 and 60,000 g/mol can only pass the membrane through pores that are larger than the respective molecule, so they will only partially permeate the membrane, which means that the rejection coefficient is between 0 and 1 ($0 < R_i < 1$). The steepness of such curves depends on the pore size distribution in a membrane.

If we transfer the MWCO of 60 kD to the graph in Fig. 2.3, we can see that this membrane belongs to the ultrafiltration (UF) range. The pore size range for UF membranes is between 2 and 200 nm. Transferred to the application in AO, such kind of *porous membranes* is suitable for the *artificial kidney*, as they largely retain albumin ($\tilde{M}_{alb} = 66.5$kD) and all proteins with higher molar mass like immunoglobulins IgG and IgM, but allow small molecules such as electrolyte ions, urea and creatinine and small proteins like ß$_2$-microglobulin to pass through. Porous UF membranes are also used for the *artificial liver*, both in the MARS system from Baxter Inc. and in the PROMETHEUS system from Fresenius (see Chap. 4.2).

Membranes with pores up to 5 μm belong to the microfiltration (MF) range. Such membranes, produced with hydrophilic polymers, are used to concentrate suspensions and are therefore suitable for *plasma donation* and *plasmapheresis*, as huge

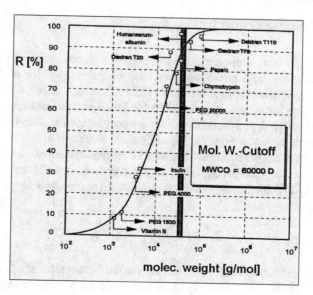

Fig. 2.2 Rejection characteristic of an ultrafiltration membrane (*Source* Melin and Rautenbach 2007). Printed with kind permission of Springer

proteins up to molar masses of 2000 kD pass through this membrane, while all blood cells are rejected. If hydrophobic polymers are used to produce MF membranes, the pores stay filled with gas and will only be wetted by plasma at rather high pressures. At the large surface of all pores on top of such membranes, which is at moderate transmembrane pressures the phase interface between blood and gas, due to gradients in partial pressures, dissolved CO_2 may desorb from blood into the gas inside the pores, while O_2 may be absorbed from gas phase into the plasma at the blood, respectively pore surface. These hydrophobic, porous MF membranes are a good choice for the *Artificial Lung* (see Chap. 4.4).

Membranes with pore sizes smaller than about 1 nm, used in nanofiltration and reverse osmosis applications, are defined to be *non-porous membranes.* Transport through such membranes may be described by the *dissolving–diffusion model* (Melin and Rautenbach 2007), in which the membrane is assumed to behave like a real liquid. Components which can be dissolved in the membrane phase are diffusively transported through the membrane along a concentration gradient and desorb into the adjacent phase on the permeate side. Those components which may not be dissolved are rejected. Since porous membranes dominate most applications in AO, the solution–diffusion model will not be discussed here.

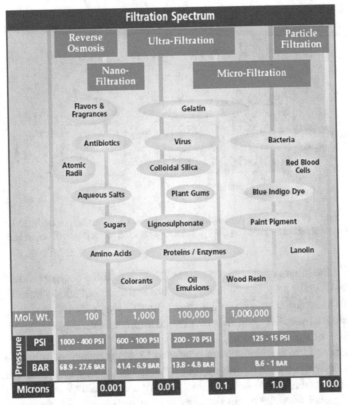

Fig. 2.3 Classification of crossflow filtration processes (*Source* GEA Membrane Filtration Brochure). Printed with kind permission of GEA

2.2 Production of Capillary Membranes and Modules

The membrane materials used in AO are predominantly organic polymers, which are optimised both in terms of transmembrane mass/volume transport and to improve biocompatibility (see references of Chap. 1).

A common production process for membranes is the phase inversion process. For experimental flat sheet membranes, a thin layer of a polymer solution is spread on a glass plate and immersed in a precipitation bath with a non-solvent for the polymer. Diffusive exchange of solvent and precipitant (non-solvent) between the two liquid phases will cause polymer to precipitate when the solubility limit is reached. The solid phase initially formed at the interface between the polymer film and the non-solvent is the resulting membrane. By varying the components (polymer, solvent, precipitant) and/or the precipitation conditions (temperature, pressure, humidity, etc.), a great variety of membrane structures can be produced.

To obtain the shape of a capillary membrane, spinnerets are used through which the highly viscous polymer solution is extruded in an annular gap and a bore liquid is

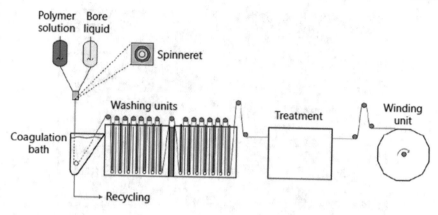

Fig. 2.4 Scheme of the spinning process to produce hollow fibre membranes by the phase inversion process (*Source* Boschetti-de-Fiero (2017)). Printed with kind permission of Baxter Inc

extruded in the lumen of the tubular polymer solution. If the polymer should already precipitate directly after the spinning jet from inside to outside as bore liquid, the precipitant has to be used. Figure 2.4 shows the spinning process up to the laying of several hundred parallel spun hollow fibre membranes into a membrane bundle in the cassettes of the winding unit.

The hollow fibre membranes are produced by an exchange of solvent and non-solvent already when both liquids meet at the outlet of the spinning jet, but also in the precipitation (coagulation) bath. In the washing units, residual solvent is removed from the hollow fibre membrane, and by further treatments like drying and undulation, the fibres are prepared for the winding unit to form membrane bundles with several thousand fibres. Undulation is a specially developed procedure to improve the bundle structure, to increase the exchange surface and to avoid fibre breakage if shrinkage occurs after potting and drying inside the housing. The winding unit is divided into equal sections, which are equipped with cassettes into which the fibres for one membrane module in a special bundle structure are placed in parallel or crossing each other. As soon as the desired number of fibres for a bundle is reached, the fibre strand is transferred to a 2nd winding wheel and the cassettes with the bundles from the 1st winding wheel can be fed to a corresponding automatic machine for the transfer into the module housings.

The further processing to get membrane modules is essentially carried out by the bundle transfer into a housing, the casting of the individual fibres to each other and to the module housing wall with an adhesive, e.g. polyurethane (PUR), the post-treatment for cleaning and for checking the tightness of membranes and modules, the labelling, the final packaging and the sterilisation (Krause 2003).

The type and number of connectors on a module housing depend on the respective application. For counter-current processes, such as dialysis and oxygenation, two connectors each are required for inlet and outlet, both on the blood side and on the dialysate or gas mixture side (see Fig. 2.1). For the crossflow filtration processes hemofiltration and plasmapheresis, one connector for the outlet from the filtrate chamber is sufficient.

The cross section of the Baxter membrane (see Fig. 2.5) is asymmetrical. If the bore liquid inside the extruded hollow fibre is the non-solvent (precipitant), phase inversion creates a 5 to 10 μm-thick layer of polymer spheres with diameters around 20–200 nm, as shown in the SEM image with the highest magnification.

From about 10 μm wall thickness, the spherical filling changes into a sponge-like structure. This structure results from the fact that the polymer starts to precipitate from the inside since the solubility limit is first reached at the contact surface between polymer solution and non-solvent. Only gradually does non-solvent diffuse through the forming polymer spherical bed into the polymer solution behind it, and thus a solid, foam-like polymer phase is formed there with a time delay. The wall of the hollow fibre shown in Fig. 2.5 is altogether about 45 μm thick, whereby the 5 to 10 μm-thick skin is decisive for the membrane characteristics. The approximately 35 μm thick sponge layer determines the mechanical strength of the fibres and thus the permissible spinning speed. The inner diameter of this membrane is about 210 μm. With this fibre diameter and an assumed effective fibre length of 20 cm, about 12,900 fibres would be necessary to get a membrane surface area of 1.7 m^2 in the module (see Eq. 2.1).

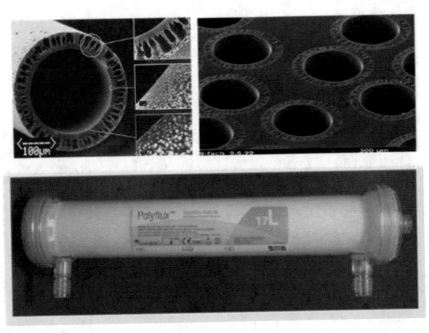

Fig. 2.5 Gambro/Baxter Dialyser Polyflux 17 L (photograph below) and SEM photographs of the membrane with different magnifications of the skin (photograph left above) and surrounded by potting material (photograph right above). Printed with kind permission of Baxter Inc

The Baxter membrane also shows an outer skin, which is more compact than the sponge structure. This is created by injecting non-solvent vapour into the spinning shaft, as the vapour condenses in contact with the outside of the extruded polymer solution tube in the spinning shaft precipitating the polymer at the still liquid outer polymer film.

In proven processes, a spinning machine with more than 200 spinnerets runs around the clock, in five shifts, seven days a week, and is only taken out of service for cleaning and preventive maintenance.

References

Boschetti-de-Fiero A, Beck W, Hildwein H, Krause B, Storr M, Zweigart C (2017) Membrane innovation in dialysis. Contrib Nephrol Basel, Karger 191:100–114. https://doi.org/10.1159/000 479259

Krause B (2003) Polymeric membranes for medical applications. CIT 75(11):1725–1732

Melin T, Rautenbach R (2007) Membrane processes. Springer, Berlin, Heidelberg

Chapter 3
Mass Transfer Models Across Membranes

Abstract Models for mass transfer across membranes are divided into those in which the membrane represents the main transport resistance and others in which additional resistances are formed by concentration boundary layers and particle layers. Since mainly porous membranes are used in AO, the pore model is derived for membrane-controlled mass transfer. Concentration boundary layers may develop if permeable components are diffusively transported along a gradient from bulk to wall concentration, or if impermeable components concentrate at the membrane surface. By establishing mass flow balances around a differential membrane cross section (a membrane element), equations are derived describing the removal rates of solvents and dissolved or dispersed molecules and its dependencies on geometrical and operational parameters. These equations will be used for the different processes in AO.

The basic operations used in process engineering can be traced back to processes through energy, momentum and mass exchange (Kraume 2012). For the applications here, energy exchange plays a subordinate role. However, mass and volume flows are essential for molecular and convective momentum and mass transfer. For these, application-specific, local relationships are derived in the following on a membrane element. In the case of a hollow fibre, this is a membrane ring with a differential length dx, which, through integration over the effective length of a hollow fibre, leads to the respective transport equations for the entire membrane surface inside a module.

Since processes with AO usually do not involve the interconnection of several membrane modules, it is sufficient to describe the performance of one module to evaluate the respective process/treatment success. To classify the processes that depend on the mode of operation, a distinction is made between models in which the transport resistances are determined on the one hand by the membrane structure and on the other hand by boundary or particle layers (Fig. 3.1).

membrane controlled mass transfer	concentration boundary-layer and particle layer controlled mass transfer
pore model *porous membranes* transport in membrane pores *dissolving-diffusion-model* *non-porous membranes* absorption of a solute from feed- into membrane phase, diffusion across membrane phase, desorption into the permeate phase	*concentration polarisation model* *crossflow-process* transport in concentration boundary-layers of impermeable components *concentration polarisation model* *counter current-process* transport in concentration boundary-layers of permeable components in feed-, membrane- and permeate-phases *particle layer model* *crossflow process* transport in particle layers

Fig. 3.1 Mass transfer models across membranes

In the case of *membrane-controlled mass transfer*, the *dissolving–diffusion model* for non-porous membranes is not considered in this book. Only the model for porous membranes used in AO today, the *pore model*, will be discussed (see Sect. 3.1). The SEM image of a porous membrane (Fig. 2.5) shows cavities (the pores) in which permeable components dissolved in a solution and the solute itself are transported from the feed to the permeate side due to concentration and pressure gradients. At a low *transmembrane pressure* (TMP) and consequently low permeate flow, in relation to the significantly higher feed flow, impermeable components and particles in the feed/retentate liquid will not accumulate on the membrane surface. In this situation, the transport resistance is mainly determined by the membrane itself.

Concentration boundary layers develop if concentrations of dissolved molecules change over the cross section of a hollow fibre membrane. This is the case if dissolved macromolecules are rejected in a *crossflow process* and concentrate at the membrane inner surface (see Sect. 3.2). In a *counter-current process*, permeable molecules will be diffusively transported from feed to permeate side if concentration gradients exist (see Sect. 3.4). Models used for these applications are summarised under *concentration boundary layer-controlled mass transfer*.

In membrane processes, where dispersed particles in suspensions are rejected by microporous membranes in a *crossflow process* (such as blood cells in plasma separation) a cake is formed on the membrane surface. Transport of solutions across this cake may be described by a similar model as the pore model, called the *particle layer model* (see Sect. 3.3) but also by a *modified boundary layer model* (see Chap. 4.2).

In all models, it is assumed that the properties of the entire membrane processed in the module remain the same regardless of location and can therefore be described at the differentially small membrane element, somewhere in the module.

3.1 Membrane-Controlled Mass Transfer: Pore Model

The approximately 5 to 10 μm-thick layer of the Baxter membrane shown in Fig. 2.5, the so-called *skin*, is the region that represents the transport resistance for the trans-membrane volume and mass flows through this type of membrane. Pores in such a membrane structure are not circular, straight channels and therefore cannot be described by simple geometric equations. For such cross sections deviating from the circular shape, equivalent diameters, such as the hydraulic diameter, are defined according to:

$$d_h = 4 \cdot A_q/U = 4 \cdot A_q \cdot \Delta z/(U \cdot \Delta z) = 4 \cdot V^g/A^s \qquad (3.1)$$

Thereby, A_q is the cross-sectional area, U the circumference and Δz the length of a pore if it is a straight hole across the membrane. V^g results from the product $A_q \cdot \Delta z$ and describes the pore volume, and the product of $U \cdot \Delta z$ is the surface area of a pore A^s. If all pores in a membrane element are regarded, V^g is the void (gas) volume resulting within the membrane element and A^s is the solid (polymer) surface of all pores. Since the real length of pores in membranes is not the membrane thickness Δz^m, an effective pore length Z^m_{por} is defined by the following equation:

$$Z^m_{por} = \mu^m \cdot \Delta z^m, \qquad (3.2)$$

where μ^m is the *tortuosity factor*, describing the "wriggled path" of a pore in a membrane. The differential area of a hollow fibre membrane element, with the inner diameter d_i of the hollow fibre and the differential length of the element dx, is calculated from:

$$dA^m = \pi \cdot d_i \cdot dx \qquad (3.3)$$

The membrane element in a hollow fibre membrane is therefore a thin, porous ring (schematic cross section of this ring (see Fig. 3.2)), through which the differential permeate volume flow $d\dot{V}^p$ may be calculated either from the product of the local flow velocity in the pores w^p_x and the differential pore cross section dA^g in the membrane element, or from the product of the volume flow density (flux) $\dot{v}^p_x$ and the inner surface of the membrane element dA^m.

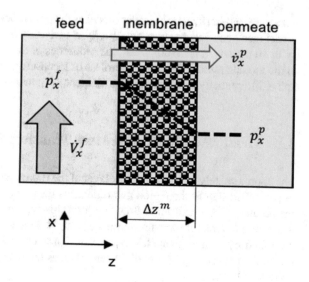

Fig. 3.2 Pore model showing pressure drop from feed to permeate, local permeate flow density (flux) $\dot{v}_x^p$ and local feed volume flow $\dot{V}_x^f$ for a differential membrane area (differential cut of the wall of a porous hollow fibre membrane)

$$d\dot{V}^p = w_x^p \cdot dA^g = \dot{v}_x^p \cdot dA^m, \tag{3.4}$$

The index x indicates the location in a hollow fibre, where the membrane element is selected. All quantities with this index are therefore local quantities for the membrane element. The quotient of the pore cross section dA^g and the inner surface of the membrane element dA^m is the surface porosity.

$$\varepsilon^m = dA^g/dA^m \tag{3.5}$$

The driving force for the differential permeate volumetric flow $d\dot{V}^p$ is the local transmembrane pressure TMP_x, which results from the difference of the local feed-side pressure p_x^f and the local permeate-side pressure p_x^p. If laminar flow is assumed, TMP_x can be calculated from d'Arcy's law and Hagen–Poiseuille's formula equations for the pressure drop in pores as follows:

$$TMP_x = p_x^f - p_x^p = \lambda_x \cdot \varrho^l/2 \cdot (w_x^p)^2 \cdot Z_{por}^m/d_h^m \tag{3.6}$$

ϱ^l is the density of the permeating liquid phase and λ_x the friction factor, which in the case of laminar pore flow may be calculated from the following equation:

$$\lambda_x = 64/Re_{Por,x} \tag{3.7}$$

The Reynolds number $Re_{Por,x}$ is a dimensionless parameter for characterising the type of flow in the pore channels and defined for these non-circular channels as follows:

$$Re_{Por,x} = w_x^p \cdot d_h^m \cdot \varrho^l/\eta^l \tag{3.8}$$

The characteristic dimension here is the hydraulic diameter d_h^m (see Eq. (3.1)). The flow velocity in the pores w_x^p calculated from Eqs. (3.4) to (3.5) gives:

$$w_x^p = d\dot{V}^p / (\varepsilon^m \cdot dA^m) = \dot{v}_x^p / \varepsilon^m \tag{3.9}$$

This results for the transmembrane pressure in:

$$\text{TMP}_x = 32 \cdot \mu^m \cdot \Delta z^m \cdot \eta^l \cdot \dot{v}_x^p / (\varepsilon^m \cdot (d_h^m)^2) \tag{3.10}$$

If all the variables characterising the membrane structure are combined to one parameter, we get the *hydraulic membrane resistance* R_h^m

$$R_h^m = 32 \cdot \mu^m \cdot \Delta z^m / (\varepsilon^m \cdot (d_h^m)^2), \tag{3.11}$$

and for the permeate volume flow density in the membrane element follows:

$$\dot{v}_x^p = \text{TMP}_x / (\eta^l \cdot R_h^m) \tag{3.12}$$

The reciprocal value from the product of dynamic viscosity η^l and hydraulic membrane resistance R_h^m is called the *hydraulic permeability of the membrane* L_P

$$L_P = 1/(\eta^l \cdot R_h^m), \tag{3.13}$$

From Eqs. (3.12) to (3.13) results a linear relationship between the permeate volume flow density (the flux) of the membrane element and the local transmembrane pressure

$$\dot{v}_x^p = L_P \cdot \text{TMP}_x, \tag{3.14}$$

The differential permeate volume flow through a membrane element can thus be calculated from:

$$d\dot{V}^P = L_P \cdot \text{TMP}_x \cdot dA^m \tag{3.15}$$

To determine the permeate volume flow for the entire module, the dependence of the transmembrane pressure on the location within the hollow fibre must be known. Assuming a linear course of transmembrane pressure as a function of x yields to the relation

$$\text{TMP}(x) = m \cdot x + b, \tag{3.16}$$

If the dynamic viscosity and the hydraulic membrane resistance, and thus the hydraulic permeability of all N hollow fibres in a module are no functions of x, the permeate volume flow for the module can be calculated from:

$$\dot{V}^P = L_P \cdot \pi \cdot N \cdot d_i \int\limits_{x=0}^{x=L_{\text{eff}}} \text{TMP}(x) \cdot dx \tag{3.17}$$

With the boundary conditions at $x = 0$; $p = p^f$ and at $x = L_{\text{eff}}$; $p = p^r$, and with the assumption that $p^p = \text{const.}$, we get for the slope m and the intercept b in Eq. (3.16) by:

$$m = -(p^f - p^r)/L_{\text{eff}}, \tag{3.18}$$

$$b = (p^f - p^p), \tag{3.19}$$

resulting in the following equation for the permeate volume flow rate in a module with porous membranes:

$$\dot{V}^P = L_P \cdot A^m \cdot \text{TMP}_m = A^m \cdot \text{TMP}_m/(\eta^l \cdot R_h^m) \tag{3.20}$$

$$TMP_m = (p^f + p^r)/2 - p^p \tag{3.21}$$

Equation (3.20) shows how permeate volume flow $\dot{V}^P$ depends via the proportionalities to $A^m \sim N \cdot d_{\text{in}} \cdot L_{\text{eff}}$ and $R^m \sim \mu^m \cdot \Delta z^m/(\varepsilon^m \cdot (d_h^m)^2)$ on geometric data of module and membrane and via $L_P \sim 1/\eta^L$ on the dynamic viscosity of the liquid used.

The following results of experiments from Eloot (2002) with the dialyser F80 from Fresenius may show the significance of the variables defined in the pore model. In Fig. 3.3, permeate volume flow ($\dot{V}^P$ = QUF) is shown as a function of the mean transmembrane pressure in the module (TMP$_m$ = Δp).

Fig. 3.3 Results of ultrafiltration experiments with the dialyser F80 (Fresenius). Ultrafiltration (QUF = UF) has been measured as a function of transmembrane pressure (TMP) using pure water before **a** and after plasma contact **b**, and with plasma **c** (*Source* Eloot 2002). Printed with kind permission of SAGE Ltd

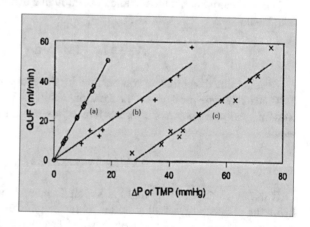

The results of the experiments with demineralised water in the sterile dialyser are represented by the measuring points around the correlation line (a). From the linear regression, the slope of this straight line (a) corresponds according to Eq. (3.20) to the product of hydraulic permeability L_P and membrane area A^m which is called the ultrafiltration coefficient:

$$\text{UFC} = L_P \cdot A^m \tag{3.22}$$

If in Eq. (3.20) the permeate volume flow rate $\dot{V}^P$ is replaced by the ultrafiltration UF, it follows:

$$\dot{V}^P = \text{UF} = \text{UFC} \cdot \text{TMP}_m, \tag{3.23}$$

For the F80 module with a membrane surface area of 1.8 m^2, the slope of line (a) in Fig. 3.3 results in an ultrafiltration coefficient for water of $\text{UFC}_{\text{H}_2\text{O}} = 2.73$ ml/(min · mmHg). From this value, the hydraulic permeability for the F80 membrane with water is:

$$L_{P,\text{H}_2\text{O}} = \text{UFC}_{\text{H}_2\text{O}}/A^M = 1.9 \cdot 10^{-10} \text{m}/(\text{s} \cdot \text{Pa}),$$

and the hydraulic membrane resistance with the dynamic viscosity of water at 37 °C ($\eta^l = 0.000693$ Pa · s) results in:

$$R_{h,}^m = 1/(L_P \cdot \eta^l) = 7.61 \cdot 10^{12} \text{m}^{-1}.$$

This order of magnitude for the transport resistance through a porous membrane is also found in classical dead-end filtration for the cake resistance of a particle layer retained by a filter element (Anlauf (2014)).

Following the measurements with water, human plasma was filtered with the same module in Eloot's experiments. The regression line (c) through these measurement points shows that the slope $\text{UFC}_{\text{Plasma}} = 1.04$ ml/(min · mmHg) of this membrane, covered with an adsorbed protein layer, is much lower than that measured with water. The correlation line (c) also no longer goes through the origin but intersects the x-axis at a pressure of about 29 mmHg. This value corresponds to the oncotic pressure of human plasma proteins that are rejected by the dialysis membrane. At transmembrane pressures below this oncotic pressure, water molecules will diffuse from permeate into feed by osmosis. In this range, filtration is consequently negative. Only when TMP exceeds the osmotic pressure difference between feed and permeate $\Delta\pi$ flux becomes positive.

After rinsing the hollow fibres with water, the measured values result in the correlation line (b), which again passes through the origin, with a slope corresponding to that of the plasma experiment. The ratio of the ultrafiltration coefficients with water before and after plasma contact gives a value of 2.63. Since manufacturers of dialysers are aware of this phenomenon, in vitro measurements to determine the UFC values for product sheets are performed after blood/plasma contact.

For the application of the pore model to solutions with molecules that are partially or completely rejected by a membrane, Eqs. (3.22) and (3.23) should be extended by the osmotic pressure difference $\Delta\pi$ between feed and permeate as follows:

$$\dot{V}^P = L_P \cdot A^m \cdot (\mathrm{TMP}_m - \Delta\pi), \tag{3.24}$$

$$\mathrm{UF} = \mathrm{UFC} \cdot (\mathrm{TMP}_m - \Delta\pi) \tag{3.25}$$

$$\Delta\pi = \pi^f - \pi^P \tag{3.26}$$

These equations for the pore model will be used for all applications in AO, in which ultrafiltration is needed (see Chaps. 4.1.1, 4.1.2, 4.2).

If Eloot would have applied in her experiments with plasma significantly higher mean transmembrane pressures, the slope of the curve measured with plasma would gradually decrease and finally reach zero. The dependence of ultrafiltration on transmembrane pressure can therefore no longer be described with the pore model.

3.2 Boundary Layer-Controlled Mass Transfer in the Crossflow Process

As mentioned in the introduction to this chapter, concentration boundary layers in the crossflow process occur when components in a solution are partially or completely rejected by the membrane. Due to a transmembrane pressure from feed to permeate side and the resulting local permeate volume flow density $\dot{v}_x^p$, such a component is transported from the bulk flow towards the membrane surface and as it is rejected, the concentration of this component at the membrane "wall" $c_{iw,x}^f$ rises above the concentration in the bulk flow $c_{ib,x}^f$ (see Fig. 3.4).

The convective transport of the regarded component i from the bulk to the wall at location x in a hollow fibre membrane can be calculated by the following equation:

$$\dot{m}_{i,\mathrm{conv},x} = \dot{v}_x^P \cdot c_{i,x} \tag{3.27}$$

The mass flow density for the diffusive back transport of component i due to the concentration gradient from the membrane surface (wall) back into the bulk (in negative z-direction and therefore the "+" in Eq. (3.28) is calculated according to Fick's 1st law:

$$\dot{m}_{i,\mathrm{diff},x} = + D_{i,j} \cdot \mathrm{d}c_{i,x}/\mathrm{d}z \tag{3.28}$$

Fig. 3.4 Model for the mass transfer in a concentration boundary layer as it occurs during crossflow filtration processes if a component i is totally rejected

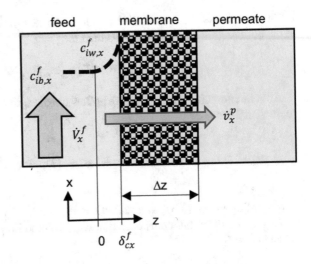

At steady-state conditions, convective transport to the wall is in equilibrium with diffusive back transport to the bulk, which means that:

$$\dot{v}_x^P \cdot c_{i,x} = D_{i,j} \cdot dc_{i,x}/dz \tag{3.29}$$

$$\dot{v}_x^P \cdot dz = D_{i,j} \cdot dc_{i,x}/c_{i,x} \tag{3.30}$$

Integrating Eq. (3.30) with the boundary conditions according to the definitions in Fig. 3.4, for $z = 0, c_{i,x} = c_{ib,x}^f$ and for $z = \delta_{c,x}$, $c_{i,x} = c_{iw,x}^f$ and assuming a constant diffusion coefficient $D_{i,j}$ the local permeate volume flow density through the membrane element is given by:

$$\dot{v}_x^P = D_{i,j}/\delta_{c,x} \cdot \ln(c_{iw,x}^f/c_{ib,x}^f) = \beta_{i,x}^f \cdot \ln(c_{iw,x}^f/c_{ib,x}^f) \tag{3.31}$$

The quotient of diffusion coefficient $D_{i,j}$ to the thickness of the concentration boundary layer $\delta_{c,x}$ is defined as *local mass transfer coefficient* $\beta_{i,x}^f$. In the Leveque solution (Leveque 1928), local mass transfer coefficient under laminar flow conditions is described by the following equation:

$$\beta_i^f(x) = D_{i,j}/\delta_{c,x}^f = 0.538 \cdot \left(D_{i,j}^2 \cdot \gamma_w/x\right)^{1/3} \tag{3.32}$$

The proportionality constant (0.538) and the exponent (1/3) were obtained by adjusting measured values from experiments. For laminar flow, wall shear rate γ_w may be calculated from the Hagen–Poiseuille equation, with the mean velocity of the feed flow w_m^f in a hollow fibre with an inner diameter d_i

$$\gamma_w = dw^f/dr_{in,r=R} = 8 \cdot w_m^f/d_i \tag{3.33}$$

The function of the local mass transfer coefficient over the length (coordinate x) of the hollow fibre membrane results then in:

$$\beta_i^f(x) = 1.076 \cdot \left(D_{i,j}^2 \cdot w_m^f/(d_i \cdot x)\right)^{1/3}, \tag{3.34}$$

After integration over the effective hollow fibre length L_{eff}, the mean mass transfer coefficient becomes:

$$\beta_{im}^f = 1/L_{\text{eff}} \cdot \int_{x=0}^{x=L_{\text{eff}}} \beta(x) \cdot dx = 1.614 \cdot \left(D_{i,j}^2 \cdot w_m^f/(d_i \cdot L_{\text{eff}})\right)^{1/3}. \tag{3.35}$$

According to Eq. (3.31) and with the assumptions of a constant wall concentration c_{iw}^f and a mean bulk concentration c_{ibm}^f, volume flow density in the module is calculated by:

$$\dot{v}^P = \beta_{im}^f \cdot ln(c_{iw}^f/c_{ibm}^f) = \beta_{im}^f \cdot lnc_{iw}^f - \beta_{im}^f \cdot lnc_{ibm}^f, \tag{3.36}$$

and the permeate volume flow in the module by:

$$\dot{V}^P = A^m \cdot \dot{v}^P = A^m \cdot 1.614 \cdot \left(D_{i,j}^2 \cdot w_m^f/(d_i \cdot L_{\text{eff}})\right)^{1/3} \cdot \ln(c_{iw}^f/c_{ibm}^f). \tag{3.37}$$

Equation (3.37) shows how permeate volume flow $\dot{V}^P$ depends on operating conditions (w_m^f), geometry of hollow fibres (d_i, L_{eff}) and physical data of the feed liquid ($D_{i,j}$, c_{ibm}^f), but no longer, as in the pore model, on the transmembrane pressure. As volume flow decreases from feed to retentate, concentration of the completely rejected component will be higher in the retentate than in the feed ($c_i^r > c_i^f$). Since concentration does not decrease linearly along a hollow fibre, a logarithmic mean is used for calculation of mean bulk concentration c_{ibm}^f

$$c_{ibm}^f = (c_i^r - c_i^f)/\ln(c_i^r/c_i^f) \tag{3.38}$$

The diffusion coefficient for macromolecules (comp. i) in a solution (comp. j) can be determined by the Stokes–Einstein equation:

$$D_{i,j} = k \cdot T/\left(6 \cdot \pi \cdot \eta_j^l \cdot r_i\right), \tag{3.39}$$

k is the Boltzmann constant, T the temperature, η_j^l the dynamic viscosity of the liquid phase and r_i the radius of the macromolecule, calculated according to Colton (1987) by the following equation:

$$r_i = (3 \cdot \widetilde{M}_i \cdot k/\left(4 \cdot \pi \cdot \tilde{R} \cdot \rho_{i,\text{hydr}}\right)^{1/3} \tag{3.40}$$

$\tilde{M}_i$ is the molar mass of the macromolecule, $\tilde{R}$ the universal gas constant and $\rho_{i,hydr}$ the density of the hydrated molecule. If the hydrated density is assumed as Colton proposed to be 1.4 g/cm³, the molecular radius results in:

$$r_i = 0.657 \tilde{M}_i^{1/3} / 10^{-10} \text{m}, \tag{3.41}$$

and the diffusion coefficient of proteins in water at 37 °C with the dynamic viscosity for water at 37 °C of $\eta^L = 0.695 \cdot 10^{-3}$Pas can be calculated from:

$$D_{i,j} = 4.97 \cdot 10^{-9} / \tilde{M}_i^{1/3} / \text{m}^2/\text{s} \tag{3.42}$$

Applying Eq. (3.37) to a crossflow process with a solution of albumin in water at constant temperature, the only variable influencing $\dot{V}^P$ would be the mean velocity in the feed w_m^f.

Comparing the results of the *pore model* with those of the *concentration polarisation model in the crossflow process*, the mass transfer resistance in the *pore model* causes a linear dependence of the volume flow density on the mean transmembrane pressure TMP$_m$, whereas in the *concentration polarisation model* the dependence on TMP$_m$ does not apply because the transfer resistance of the fully developed boundary layer dominates the transport.

The transitions between both models were investigated in experiments by Neggaz (2007) with an aqueous albumin solution. The results are shown in Figs. 3.5 and 3.6. If the permeate volume flow density $\dot{v}^P = $ Jf is measured up to high transmembrane pressures, the course of the experimental points can be divided into three regions. In the example of Fig. 3.5 up to about 1 bar, there is a linear relationship between Jf and TMP$_m$. In this range, the equations of the pore model may be applied. The hydraulic membrane resistance R_h^m dominates the mass transport.

From a TMP$_m$ of about 3 bar, the volume flow density remains constant, and a plateau is reached at a flux of about $1.4 \cdot 10^{-5}$m/s (see Fig. 3.5). The concentration

Fig. 3.5 Results of crossflow ultrafiltration experiments with an albumin–water solution. Permeate volume flow density (flux, $\dot{v}^P = $ Jf) is shown as a function of the mean transmembrane pressure (TMP$_m$) (*Source* Neggaz (2007)). Printed with kind permission of Elsevier

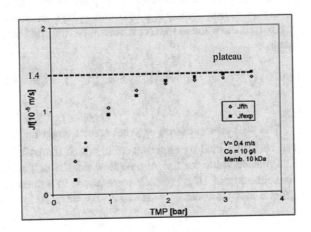

Fig. 3.6 Permeate volume flow density (flux J) as a function of mean albumin concentration (Co = cbm) in crossflow ultrafiltration of albumin–water solutions (*Source* Neggaz (2007)). Printed with kind permission of Elsevier

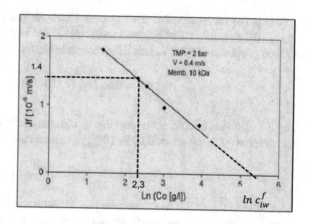

boundary layer is completely developed, and according to the model described in this chapter, the feed-side boundary layer resistance $R_i^f = 1/\beta_{im}^f$ dominates the mass transfer. According to Eq. (3.37), the achievable plateau of the volumetric flow density may not be further increased by a higher TMP_m. A higher mean velocity w_m^f however would reduce the boundary layer thickness, thus increasing the mass transfer coefficient β_{im}^f and the volume flow density (see Chap. 4.1.1). In the transition range $1 < TMP_m < 3$ bar, the gradient $d\dot{v}^P/dTMP_m$ decreases continuously and becomes zero in the plateau region.

According to Neggaz (2007), by combining the pore and the boundary layer model, the dependence shown in Fig. 3.5 for the entire TMP range can be described with the following equation:

$$\dot{v}^P = x_{cr} \cdot TMP/(L_{eff} \cdot \eta \cdot R^m) + 1,614 \cdot \left(D_{i,j}^2 \cdot w_m^f/(d_i \cdot L_{eff})\right)^{1/3} \cdot$$
$$\ln(c_{iw}^f/c_{ibm}^f) \cdot \left((L_{eff}^{2/3} - x_{krit}^{2/3})/L_{eff}\right) \tag{3.43}$$

The length x_{cr} is defined to be the place inside a hollow fibre bundle, where the local volume flow density $\dot{v}_{xcr}^P$ for boundary layer-controlled mass transfer (Eq. (3.44), left part) is the same as that for membrane-controlled mass transfer (Eq. (3.44), right part).

$$1.076 \cdot \left(D_{i,j}^2 \cdot w_m^f/(d_i \cdot x_{cr})\right)^{1/3} \cdot \ln(c_{iw}^f/c_{ibm}^f) = (p^f - 32 \cdot \eta \cdot w_m^f \cdot x_{cr}/D_{i,j}^2)/\eta \cdot R^m \tag{3.44}$$

The wall concentration c_{iw}^f that results when the boundary layer is developed can be determined by experiments in which the bulk concentration c_{ibm}^f is varied. The plateau value for the volume flow density achieved in Fig. 3.5 at an albumin concentration of 10 g/l (which corresponds to a numerical value of 2.3 on a logarithmic scale) is about $\dot{v}^P = Jf = 1.4 \cdot 10^{-5}$m/s. At lower concentrations $c_{ibm}^f = C_0$, the volume flow density increases according to Eq. (3.37), and a higher plateau is reached.

At concentrations higher than 10 g/l, the plateau value decreases along the correlation line in Fig. 3.6 and becomes $\dot{v}^P = Jf = 0$ m/s if $c_{ibm}^f = c_{iw}^f$. Extending the straight line to the point of intersection with the x-axis, one finds there the wall concentration with a value of $c_{iw}^f = e^{5.28} = 196$ g/l. Thus, in the example considered here, the bulk concentration of 10 g/l will be increased at a mean velocity of 0.4 m/s by a factor of 19.6 to get the wall concentration. The negative slope of the regression line in the semi-logarithmic representation according to Eq. (3.36) gives the mean mass transfer coefficient β_{im}^f.

3.3 Particle Layer-Controlled Mass Transfer

The filtration of suspensions is often carried out in a dead-end process, whereby the rejected particle layer becomes thicker over time and the permeate volume flow decreases quickly, despite readjustment of the pressure. The speed with which a maximum pressure drop caused by the cake occurs determines the duration of the filtration interval. At the latest, when the pressure drop is as high as the maximum permissible transmembrane pressure for the membrane or the module, filtration must be interrupted and the filter medium (the membrane) must be backwashed. This discontinuous mode of operation can be circumvented by using the crossflow method instead of the dead-end method. In this case, a cake will still build up in front of the membrane, but it can be adjusted by choosing suitable operating parameters so that the pressure drop does not reach the permissible transmembrane pressure.

According to a proposal by Rippberger (1993), for the calculation of the permeate volume flow in the crossflow microfiltration of suspensions, analogous to the hydraulic membrane resistance R_h^m a particle layer resistance R^{pl} has been introduced. Since a filter cake can be described analogously to a membrane skin of polymer beads as a particle layer with the porosity ε^{pl}, tortuosity μ^{pl}, hydraulic diameter d_h^{pl} and thickness Δz^{pl}, it follows analogously to the hydraulic membrane resistance R_h^m (see Eq. (3.11) for the particle layer resistance):

$$R^{pl} = 32 \cdot \mu^{pl} \cdot \Delta z^{pl} / (\varepsilon^{pl} \cdot (d_h^{pl})^2) \tag{3.45}$$

Assuming the two resistances add up during crossflow microfiltration of a suspension, the permeate volume flow according to the particle layer model is calculated by:

$$\dot{V}^P = A^m \cdot TMP_m / [\eta^l \cdot (R_h^m + R^{pl})] \tag{3.46}$$

The characteristic quantities for the surface layer according to Eq. (3.45) can be determined from a particle size analysis and from filtration experiments. By rearranging Eq. (3.46), the following is obtained for the particle layer resistance

$$R^{pl} = A^m \cdot TMP_m / \left(\dot{V}^p \cdot \eta^l \right) - R_h^m \tag{3.47}$$

Since the simple model of particle-layer-controlled mass transfer presented here cannot establish a connection between the target variable $\dot{V}^p = $ UF and geometric and operational variables, extensions of the model resulted, such as that of Zydney (1982), which will be discussed in Chap. 4.2, Plasma Donation and Plasma Exchange.

3.4 Boundary Layer-Controlled Mass Transfer in the Counter-Current Process

3.4.1 Mass Transfer Between Liquid Phases

It is known from technical applications that counter-current flow in heat exchangers is more effective for energy transfer than co-current flow. Since there is an analogy between heat and mass transfer under comparable flow conditions, the equations for mass transfer in the counter-current process can be derived analogously to those for heat transfer, if temperature gradients are replaced by concentration gradients (Bird 2007).

The model for boundary layer-controlled mass transfer in a counter-current process is again developed for a differential membrane surface, the membrane element, at any point x in the hollow fibre membrane (see Fig. 3.7).

Assuming that there is no ultrafiltration through the membrane element (TMP$_m$ = 0), dissolved permeable components are transported through the membrane exclusively by diffusion. This transport takes place from feed to permeate through three

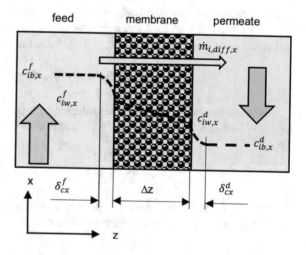

Fig. 3.7 Model for the boundary layer and membrane phase-controlled mass transfer in counter-current processes. Dissolved component i is assumed to be totally permeable

zones, within each of which a concentration gradient in the positive z-direction will be adjusted. The concentration of the considered permeable component i is highest in the bulk flow of the feed side. The concentration at the feed-side membrane surface $c_{iw,x}^{f}$ is lower than that in the bulk feed flow $c_{ib,x}^{f}$, and the region between these two concentrations is the feed-side boundary layer δ_{cx}^{f}. The concentration further decreases through the membrane pores and reaches on the outer surface of the hollow fibre membrane the dialysate wall concentration $c_{iw,x}^{d}$. In the dialysate boundary layer concentration is further decreasing from the wall to the dialysate bulk concentration $c_{ib,x}^{d}$.

Under steady-state condition for each region, Fick's 1st law may be used to describe the diffusive mass transport of component i.

$$\dot{m}_i = -D_{i,j} \cdot dc_i/dz \qquad (3.48)$$

After integration, the following mass flow densities through a membrane element inside a hollow fibre result at location x:

Feed-side boundary layer

$$\dot{m}_{i.\text{diff},x}^{f} = \beta_{i,x}^{f} \cdot (c_{ib,x}^{f} - c_{iw,x}^{f}) \qquad (3.49)$$

Membrane

$$\dot{m}_{i.\text{diff},x}^{m} = P_i^{m} \cdot (c_{iw,x}^{f} - c_{iw,x}^{d}) \qquad (3.50)$$

Dialysate-side boundary layer

$$\dot{m}_{i.\text{diff},x}^{d} = \beta_{i,x}^{d} \cdot (c_{iw,x}^{d} - c_{ib,x}^{d}) \qquad (3.51)$$

The mass transfer coefficient in the membrane P_i^{m} is assumed to be a constant value for all membrane elements in the module. It is called the *diffusive membrane permeability* for component i and can be calculated analogously to the mass transfer coefficients in the boundary layers from the quotient of the diffusion coefficient of component i in the membrane pores D_i^{M} and the wall thickness of the hollow fibre Δz.

$$P_i^{m} = D_i^{m}/\Delta z \qquad (3.52)$$

Since the wall concentrations are difficult to measure, an *overall mass transfer coefficient* K_{0i} is defined for the transport through all three regions, and thus, the following equation for the mass flow density through the membrane element is established:

$$\dot{m}_{i,diff,x} = K_{0i,x} \cdot (c_{ib,x}^{f} - c_{ib,x}^{d}) \qquad (3.53)$$

Since in steady-state condition all mass flow densities must be equal, the following relationship between the *local overall mass transfer coefficient* $K_{0i,x}$, the *local mass transfer coefficients in feed and dialysate boundary layers* β_{ix}^f, β_{ix}^d and the *diffusive membrane permeability* P_i^m is derived from Eqs. (3.49) to (3.53):

$$1/K_{0i,x} = 1/\beta_{ix}^f + 1/P_i^m + 1/\beta_{ix}^d \tag{3.54}$$

After integration of the local relationships derived for a membrane element at location x, the diffusive mass flow of a permeating component i results in:

$$\dot{M}_{i,diff} = \dot{m}_{i,diff} \cdot A^m = K_{0im} \cdot A^m \cdot \Delta c_{im}, \tag{3.55}$$

where K_{0im}, the mean overall mass transfer coefficient of component i in the regarded module, is calculated from:

$$1/K_{0im} = 1/\beta_{im}^f + 1/P_i^m + 1/\beta_{im}^d \tag{3.56}$$

The mean feed-side mass transfer coefficient β_{im}^f can be calculated from Eq. (3.35). The mean dialysate-side mass transfer coefficient β_{im}^d will depend on the bundle structure in the housing, and therefore, a relationship analogous to the feed side must be determined experimentally for each module design (Chap. 4.1.1). Diffusive membrane permeability P_i^m is often measured for relevant components (like urea for dialysis membranes) by the membrane manufacturer.

If the reciprocal values of the mass transfer coefficients are defined to be *mass transfer resistances*, it follows

$$1/K_{0im} = R_{i,tot} = R_i^f + R_i^m + R_i^d \tag{3.57}$$

Then, *total mass transfer resistance in a membrane module* $R_{i,tot}$ is the sum of the resistances for the diffusive transport in the boundary layers and in the membrane pores.

The driving force for the diffusive transport in the membrane module is the mean logarithmic concentration difference, given by:

$$\Delta c_{im} = (\Delta c_{i\alpha} - \Delta c_{i\omega})/\ln(\Delta c_{i\alpha}/\Delta c_{i\omega}) \tag{3.58}$$

The logarithmic mean value of the concentration difference Δc_{im} is calculated from the local concentration differences at the feed-side inlet (= dialysate-side outlet in a counter-current process).

$$\Delta c_{i\alpha} = c_{i,in}^f - c_{i,out}^d \tag{3.59}$$

and at the feed-side outlet (= dialysate-side inlet).

$$\Delta c_{i\omega} = c_{i,out}^f - c_{i,in}^d \qquad (3.60)$$

How diffusive mass flow $\dot{M}_{i,diff}$ of permeable components depends on geometrical, operational parameters can be derived from Eqs. (3.55) to (3.60) and will be discussed in detail for the removal rates of components as it should be eliminated by dialysis in an *Artificial Kidney* (see Chap. 4.1.1).

By setting an average transmembrane pressure in the module, permeable components dissolved in the feed are, together with the solvent, convectively transported across the membrane. At steady-state condition, the balance for component i is calculated from:

$$\dot{M}_i = \dot{V}_{in}^f \cdot c_{i,in}^f - \dot{V}_{out}^f \cdot c_{i,out}^f \qquad (3.61)$$

With $\dot{V}^P = \dot{V}_{in}^f - \dot{V}_{out}^f$, it follows:

$$\dot{M}_i = \dot{V}_{in}^f \cdot c_{i,in}^f - \left(\dot{V}_{in}^f - \dot{V}^P\right) \cdot c_{i,out}^f \qquad (3.62)$$

$$\dot{M}_i = \dot{V}_{in}^f \cdot (c_{i,in}^f - c_{i,out}^f) + \dot{V}^P \cdot c_{i,out}^f \qquad (3.63)$$

In contrast to purely diffusive mass transfer ($\dot{V}^P = 0\,ml/min$), the mixed diffusive and convective transport is by $\dot{V}^P \cdot c_{i,out}^f$ higher (see Chaps. 4.1.1 and 4.1.2).

3.4.2 Mass Transfer Between Gas and Liquid Phases

In contrast to processes, in which an exchange of components between two miscible liquid phases takes place, as in dialysis, in this chapter the exchange between components across a gas–liquid phase interface will be discussed. The condition for achieving an equilibrium between gas and liquid phases is that the partial pressures of the transferred component i in the adjacent phases are equal.

For an ideal gas mixture, at the pressure p the relationship between the molar fraction of a component i in the gas phase $\tilde{y}_i$ and the partial pressure of this component in the gas p_i^g is described by *Dalton's law*

$$p_i^g = \tilde{y}_i \cdot p \qquad (3.64)$$

For the partial pressure of component i in a real liquid phase p_i^L, *Henry's law* may be applied for small molar fractions of component i in the liquid phase $\tilde{x}_i$ as follows:

$$p_i^l = \tilde{x}_i \cdot He_i(T) \qquad (3.65)$$

$He_i(T)$ is the temperature-dependent Henry constant. From the two Eqs. 3.64 and 3.65 follows for the relation between the molar fractions of a component i in phase equilibrium ($p_i^g = p_i^l$):

$$\tilde{x}_i = \tilde{y}_i \cdot p/He_i(T) \tag{3.66}$$

If, for example, oxygen is dissolved from ambient air at 20 °C and 1 bar in water, with the Henry coefficient $He_{O2}(20°C) = 40537$ bar, and the molar fraction of O_2 in air is $\tilde{y}_{O2} = 0.21$ mol O_2/mol air, the molar fraction of oxygen in water becomes $\tilde{x}_{O2} = 5.18 \cdot 10^{-6}$ mol O_2/mol water. If this concentration, which is about five orders of magnitude smaller than the molar O_2 fraction in air, is reached in water, both phases are with respect to O_2 in equilibrium which means that O_2 concentrations in both phases will not change anymore. In phase equilibrium between gas and liquid phases, therefore concentrations of a component could be quite different. To avoid this discontinuity in the concentration curve over the gas–liquid phase interface, partial pressures are used instead.

In Fig. 3.8, it is assumed that the pores of the microporous, hydrophobic membrane are not wetted with liquid and that the gas phase fills the pore space. The exchange area (phase interface) between gas and liquid phases thus results from the surface area of all membrane pores facing the liquid phase.

The driving force for the diffusive mass transport of a component i through the liquid-side boundary layer is the local gradient between the volume concentration at the membrane surface $c_{iw,x}^{lv}$ and the volume concentration in the bulk flow $c_{ib,x}^{lv}$ which changes over the length of the liquid flow channel in the module (coordinate x). Assuming a constant gas-side partial pressure, the equilibrium concentration in the liquid phase resulting at this partial pressure will not change with x ($c_{iw,x}^{lv} = c_{i,w}^{lv}$). The local concentration differences in the liquid boundary layer along the

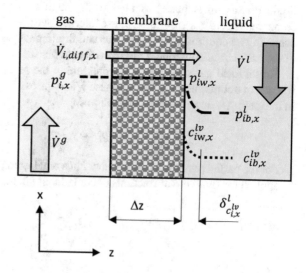

Fig. 3.8 Model for the transport of component i from gas to liquid phase in a membrane element with a porous, hydrophobic membrane (counter-current process)

flow channel are therefore between $\Delta c_{i\alpha} = c_{i,w}^{lv} - c_{ib,in}^{lv}$ at the inlet to the module and $\Delta c_{i\omega} = c_{i,w}^{lv} - c_{ib,out}^{lv}$ at the exit from the module. For the mean, logarithmic concentration difference across the module, it thus follows:

$$\Delta c_{i,m}^{lv} = (c_{ib,out}^{lv} - c_{ib,in}^{lv})/\ln((c_{i,w}^{lv} - c_{ib,in}^{lv})/(c_{i,w}^{lv} - c_{ib,out}^{lv})), \tag{3.67}$$

The diffusive volume flow of a component i in the counter-current process can be described equivalent to Eq. (3.55) by using the volume concentration $\Delta c_{i,m}^{lv}$ instead of the mass concentration $\Delta c_{i,m}^{l}$.

$$\dot{V}_{i,diff} = K_{0im} \cdot A^m \cdot \Delta c_{i,m}^{lv} \tag{3.68}$$

When using microporous, hydrophobic membranes, the overall mass transfer coefficient K_{0im} may be replaced by the mass transfer coefficient in the liquid phase $\beta_{i,m}^L$ since the mass transfer resistances in the gas and membrane phases are negligibly small.

$$K_{0,i,m} = \beta_{i,m}^l \tag{3.69}$$

Thus, the diffusive volume flow of a component i $\dot{V}_{i,diff}$ from the gas side to the liquid results on the one hand from the equation for the mass transfer in the liquid film and on the other hand from the balance for the considered component i on the liquid side in steady-state operation:

$$\dot{V}_{i,diff} = \beta_{i,m}^l \cdot A^m \cdot \Delta c_{i,m}^{lv} = \dot{V}^l \cdot (c_{ib,out}^{lv} - c_{ib,in}^{lv}) \tag{3.70}$$

If the two phases flow in counter-current to each other and the liquid flow is laminar, the mean mass transfer coefficient of component i in the liquid β_{im}^l can be calculated according to Leveque's approach (Eq. 3.35). This results in the following dependencies of the diffusive volume flow of a component i through the membrane on geometric, operating, and physical data.

$$\dot{V}_{i,diff} = 1.614 \cdot (D_{i,j}^2 \cdot w_m^l/(d_i \cdot L_{eff}))^{1/3} \cdot A^m \cdot \Delta c_{i,m}^{lv} \tag{3.71}$$

If components should be eliminated from the liquid into the gas phase, the same equations may be used; however, the concentration gradient will decrease in the negative z-direction, because the bulk concentration of such components must be higher than the wall concentration.

References

Anlauf H (2014) Solid-liquid separation through cake filtration. Chemical Weekly, 207–213

Bird RB, Stewart WE, Lightfood EN (2007) Transport phenomena, Revised 2nd edn, Wiley, ISBN: 978–0–470–11539–8

Colton CK (1987) Analysis of membrane processes for blood purification. Blood Purif 5:202–251

Eloot S, De Wachter D, Vienken J, Pohlmeier R, Verdonck P (2002) In vitro evaluation of the hydraulic permeability of polysulfone dialysers. Intern J Art Org 26(2):210–218

Kraume M (2012) Transportvorgänge in der Verfahrenstechnik. Springer

Lévêque A (1928) Les lois de la transmission de chaleur par convection Ann. Mines 13:201–299, 305–362, 381–415

Neggaz Y, Lopez Vargas M, Ould Dris A, Riera F, Alvarez R (2007) A combination of serial resistances and concentration polarization models along the membrane in ultrafiltration of pectin and albumin solutions. Separation and Purification Techn 54:18–27

Rippberger S (1993) Berechnungsansätze zur Mikrofiltration, Chem Ing Tech 65 (1993) Nr. 5, S. 533–540

Zydney AL, Coloton CK (1982) Continuous flow membrane plasmapheresis: Theoretical models for flux amd hemolysis prediction. Trans Am Soc Art Intern Organs 28:408–412

Chapter 4
Membrane Processes in Artificial Organs

Abstract In Artificial Organs, different, life-saving organ-specific requirements must be realised. Artificial Kidney involves the removal of water-soluble components with molecular masses up to about 20kD. In dialysis, the healthy concentrations are adjusted by exchange with an adequate dialysate; in hemofiltration, this is regulated by substituting appropriate solutions. In the Artificial Liver, hydrophobic toxins to be removed are bound to albumin (62kD). In the MARS, the dissociated part of the toxin diffuses through an ultrafiltration membrane, binds in a dialysate circuit to human albumin and is separated from albumin by adsorption and ion exchange. In the PROMETHEUS system, albumin-bound toxins (ABTs) are removed by a micro-porous membrane. After removal of the toxin by adsorption, patient's albumin is returned to blood by back filtration. An Artificial Lung exchanges the gases O_2 and CO_2, which are physically dissolved in plasma but also chemically bound to blood components. The driving force for the transfer is gradients in the partial pressures of O_2 and CO_2. During Plasma Exchange, components with a molar mass up to 1000kD are separated from blood cells and plasma loss will be substituted.

The medical application of membranes in Artificial Organs presupposes that in the case of acute and/or chronic insufficiency of an organ, the vital functions can be taken over by the Artificial Organ. In this case, an extracorporeal circuit is set up between the patient and the Artificial Organ via accesses to blood lines, in which suitable "hardware" (pumps, measuring devices, MSR components, etc.) ensures safe circulation of the blood (see Fig. 4.1, blood monitor).

For the very different, organ-specific requirements, individual adaptations of membrane, module properties and processes are necessary. For example, controlled removal of enriched, water-soluble components in haemodialysis is achieved by dosing the valuable blood solutes in a fluid circuit by mixing water with a concentrate in such a way that the dialysate at the entrance to the module contains these solutes in the physiologically "normal" concentration. When these concentrations are reached in the blood during the treatment, the concentration gradient between blood and dialysate disappears and thus the diffusive transport of substances ends. The correct composition in the fluid circuit is controlled by measuring the conductivity,

M. Raff, *Mass Transfer Models in Membrane Processes*,
SpringerBriefs in Bioengineering,
https://doi.org/10.1007/978-3-030-89195-4_4

blood monitor

Basic Functions
-Preparation and circulation of the liquids
 (blood, heparin, dialysate)
-Mass balance

Security Functions
monitoring of:
- pressures,
- airbubbles, blood
- blood leakage, dialysate.
- composition of the dialysate

Auxiliary functions
adaptation of
- single needle technology
- Highflux-Dialysis

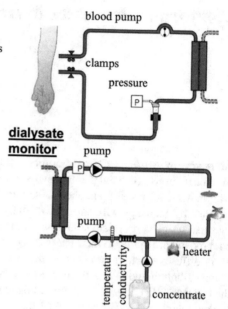

Fig. 4.1 Important functions for blood and dialysate monitors in dialysis. Printed with kind permission of Baxter Inc

and the solution is heated to body temperature, degassed, and pumped. Regulation of the pressures and pump speeds ensures the necessary transmembrane pressure for fluid removal (see Fig. 4.1, dialysate monitor).

Non-water-soluble, hydrophobic components, such as bilirubin, bile acid and some active components of drugs, are transported in the plasma by binding to proteins. Since the proteins are to be preserved in the blood, the hydrophobic component must be separated from the carrier protein in the *Artificial Liver* and excreted, while the protein should remain in the patient's bloodstream. For *Artificial Kidney* and *Artificial Liver*, counter-current as well as crossflow processes are applied.

In the case of counter-current processes dialysis (see Sect. 4.1.1) and MARS (see Sect. 4.3), the transport is dominated by diffusion and therefore determined by the composition of the receiving phases beyond the membrane as shown in Fig. 4.1.

While in *Artificial Kidney* and *Artificial Liver* components are exchanged between two liquid phases via a membrane, in the *Artificial Lung* the exchange takes place via a gas–liquid phase interface (see Sect. 4.4). Membrane oxygenators of the first generation also used the counter-current principle as known from dialysis, but soon it was recognised that module designs, where blood flows outside of hollow fibres perpendicular to the gas inside the fibres, result in much better performances. In the crossflow processes Hemofiltration (see Sect. 4.1.2), Plasma Donation and Plasma

Exchange (see Sect. 4.2) and PROMETHEUS (see Sect. 4.3), the ratio of filtration and substitution fluid is regulated by volume/mass balances (see, e.g., Fig. 4.8).

4.1 Artificial Kidney

The functions of the human kidney can be roughly divided into excretoric (cleansing) and secretoric (regulating) functions (Fig. 4.2).

Only the *excretoric functions* can be taken over by an *Artificial Kidney*. Waste products are uraemic toxins that may totally be excreted if they can pass through the membrane. The elimination of water in the order of 2 L per treatment is regulated by controlled ultrafiltration via the set transmembrane pressure. Since dialysis membranes are permeable to electrolytes, these would be completely discharged without appropriate measures, which would lead to fatal deficiency symptoms. A drop in electrolytes below the physiological healthy level must consequently be prevented by appropriate addition of electrolytes to the dialysate in counter-current dialysis or by substitution of a sterile electrolyte solution in hemofiltration. Since kidney patients suffer from hyperacidity of the blood, the pH value in the blood is regulated by dosing acetate or bicarbonate buffer into the dialysate.

For the *secretoric functions*, appropriate drugs must be administered during or between treatments. This administration of medication becomes necessary when the functions of the adrenal cortex are also lost in kidney failure. Then, e.g. the hormone erythropoietin (EPO) is no longer produced, which gradually reduces the number of red blood cells in the patient's blood. In the 1970s and 1980s, kidney patients therefore still received transfusions of donor blood or red blood cell concentrate at regular intervals so that the haematocrit did not fall below 20%. Since this value is significantly below the normal value of healthy people (40–45%), the patients'

Fig. 4.2 Excretoric and secretoric functions of human kidney. Printed with kind permission of Baxter Inc

Excretoric kidney functions

Elimination of waste

Elimination of water

Regulation of acid - base balance

Regulation of electrolyte concentration

Secretoric kidney functions

Regulation of blood pressure (renin)

Regulation of red blood cells (EPO)

Regulation of calcium (vitamin D)

performance was considerably impaired. It was not until 1989, when a recombinant EPO preparation came on the market, that the haematocrit of patients could be permanently raised to about 30% by administering this preparation accordingly. In today's treatments, the administration of EPO is standard.

After the dialysis-pioneers Kolff and Higgins (1954) and Alwall et al. (1948) were able to show with their treatments that with membranes and suitable processes it was possible to fulfil the excretory kidney functions extracorporeally and thus save human lives, various companies began to develop the process further. The medical doctor Nils Alwall advised the Swedish entrepreneur Holger Crafoord in his efforts to build up the later company Gambro (today Baxter), which, like others (Fresenius, Braun, Hospal, Asahi, etc.), set itself the goal of developing, producing and selling the necessary components and machines for this treatment.

Meanwhile, different processes compete for the best outcome for the patient. The preferred solution, if a suitable donor organ is available, is transplantation. Patients who want to be largely independent of equipment use their peritoneum as a membrane and fill and empty the abdominal cavity at intervals of a few hours. Since peritoneal dialysis uses the body's own organ and not an artificial one, this topic will not be further discussed here.

For treatments in dialysis units or at home, the procedures haemodialysis and hemofiltration are performed. The advantages of treatments on the ward are the expert care of the patients and the monitors needed for these treatments. If there is a suitable family environment, treatment at home (home dialysis) is also possible after appropriate training. On the ward, patients are usually treated two to three times a week for 3–5 h at fixed appointments. Home dialysis patients can choose their treatment times flexibly also overnight.

In the case of chronic kidney disease (ESRD = end-stage renal disease), the patient's access to the extracorporeal blood circulation is via a so-called shunt. In this case, an artery is connected to a vein through a surgical procedure. The higher arterial pressure leads to an expansion of the vein, so that on the one hand the needles can be placed well, and on the other hand significantly higher extracorporeal blood flows are possible. While in blood donation the flows are only about 30–50 ml/min, in haemodialysis blood flows of 200–400 ml/min are common, and in hemofiltration up to 500 ml/min.

4.1.1 Dialysis

Dialysis treatments were initially carried out with so-called *Lowflux* membranes. For such membranes, manufacturing conditions had been developed that resulted in ultrafiltration coefficients of about UFC = 5ml/($h \cdot$ mmHg). If the amount of liquid to be removed from a patient over four hours treatment time is assumed to be two litres, an ultrafiltration of UF = 8.3ml/min. would be necessary. This flow perpendicular to the membrane surface is significantly lower than the usual blood flows in dialysis of 200 to 400 ml/min. Therefore, in *Lowflux Dialysis (LFD)* it is assumed that no concentration boundary layer of rejected proteins will develop on the membrane surface and the main transport resistance will be the membrane. The dependence of ultrafiltration on the mean transmembrane pressure may then be described with the equation for the *pore model* (see Chap. 3.1, Eq. (3.25)).

$$UF = UFC \cdot (TMP_m - \Delta\pi) \tag{4.1}$$

With an osmotic pressure difference of $\Delta\pi = 30mmHg$ and an ultrafiltration coefficient of $UFC = 5ml/(h \cdot mmHg)$, mean transmembrane pressure should be $TMP_m = 130mmHg$ to get an ultrafiltration of $UF = 500ml/h$. In LFD, local transmembrane pressures are always positive, which means that local UF along the fibres is always a filtration from blood to dialysate side.

Highflux membranes had been developed because long-standing dialysis patients complained of pain in their joints. The cause was found to be deposits, amyloid fibrils, consisting mainly of ß$_2$-microglobulin (ß$_2$-M). This protein has a molecular weight of about 12,000 g/mol and is largely rejected by *Lowflux* membranes. As a result, its concentration in the patient's plasma increases continuously and leads to haemodialysis-associated amyloidosis. This is avoided if membranes with a sieving coefficient for ß$_2$-M of about 0.7–0.8 are used. However, increasing the permeability of membrane pores for the elimination of large molecules has the consequence that also ultrafiltration increases. In treatments where *Highflux* membranes with an ultra-filtration coefficient of $UFC = 55ml/(h \cdot mmHg)$ are used, mean transmembrane pressure must be reduced to a value of about $TMP_m = 39mmHg$ to keep as in LFD the ultrafiltration at $UF = 500ml/h$. With such a low TMP_m, the local TMP_x along a hollow fibre membrane may become negative. In counter-current processes, TMP_x results from the difference of the local pressures on the blood and dialysate side:

$$TMP_x = p_x^b - p_x^d, \tag{4.2}$$

Assuming in HFD electrolyte concentrations on blood and dialysate sides balance out, local osmotic pressure difference at the membrane element is calculated solely from the osmotic pressure of the macromolecules (proteins) rejected on the blood side

$$\Delta\pi_x = \pi_{Makro,x} \tag{4.3}$$

Then, local effective transmembrane pressure is defined by:

$$TMP_{eff,x} = TMP_x - \pi_{Makro,x} \tag{4.4}$$

and local UF_x becomes zero if $TMP_x = \pi_{Makro,x}$. Calculations by Raff (2003), applied to *HFD* with an aqueous dextran solution as "blood substitute" and water as dialysate, show the dependencies of local flows and local pressures for the Polyflux 210H *Highflux* dialyser from Baxter over the hollow fibre length (coordinate x). The left diagram in Fig. 4.3 shows that ultrafiltration gradually increases to 20 ml/min

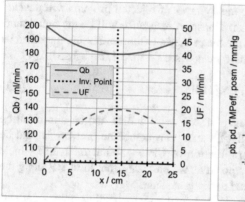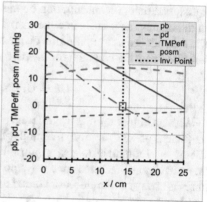

Fig. 4.3 Flow of dextran–water solution on blood side (Qb), ultrafiltration (UF) (left diagram) and pressures (right diagram) as functions of fibre length in a *Highflux* dialyser (Polyflux 210H, Baxter/Gambro) for a dialysis experiment with a dextran–water solution (inlet flow: 200 ml/min; inlet dialysate flow: 500 ml/min). Printed with kind permission of Baxter Inc

up to a hollow fibre length of approximately $x = 14$ cm and decreases to the total UF of 10 ml/min at the end of the dialyser. Accordingly, dextran flow decreases from 200 ml/min to 180 ml/min and then increases to 190 ml/min. This can be explained by the fact that the effective transmembrane pressure in the front area of the dialyser ($0 < x < 14$ cm) is positive (right diagram in Fig. 4.3) and causes a *forward filtration* from the dextran side to the dialysate side. In the rear area of the dialyser (14 cm $< x <$ 25 cm), $\text{TMP}_{\text{eff},x}$ becomes negative, with the consequence that a *back filtration* from the dialysate side to the dextran side occurs in this region of the dialyser. The "inversion point line" is at length x where $\text{TMP}_{\text{eff},x}$ becomes zero (dotted black line and small red square). The osmotic pressure on the dextran side $\pi_{\text{Makro},x} = p_{\text{osm}}$ increases up to the "inversion point" because the concentration of the impermeable macromolecule dextran increases as water is ultrafiltered towards the dialysate. From the "inversion point" to the end of the dialyser ($x = L_{\text{eff}}$), the osmotic pressure decreases due to the dilution of the dextran solution caused by the *back filtration* of water from dialysate to dextran side.

From these experiments, one may conclude that in *Highflux* dialysis, *back filtration of* non-sterile dialysate into the patient's blood is possible and therefore it must be ensured that endotoxins present in the dialysate are safely rejected over the treatment period. This is achieved by taking appropriate measures during membrane production, for example, by moderate precipitation of the polymer solution from the outside (Chap. 2) and by selecting suitable membrane polymers that may adsorb endotoxins.

Together with the ultrafiltrate which mainly regulates the water balance in patients, small amounts of dissolved uraemic toxins are removed convectively. Total removal rate of dissolved permeable components may be calculated by the following modified Eq. (3.63), in which $\dot{V}_{\text{in}}^f$ is replaced by Q_{in}^b and $\dot{V}^p$ by *UF*:

$$\dot{M}_i = Q_{\text{in}}^b \cdot \left(c_{i,\text{in}}^b - c_{i,\text{out}}^b\right) + \text{UF} \cdot c_{i,\text{out}}^b \qquad (4.5)$$

If this mass flow $\dot{M}_i$ is related to the entry concentration of a regarded component i into the dialyser $c_{i,\text{in}}^b$, the result is the *clearance of component i*, a quantity that describes the proportion of the inlet blood flow Q_{in}^b that is completely freed from this component:

$$\text{CL}_i = \dot{M}_i / c_{i,\text{in}}^b = Q_{\text{in}}^b \cdot \left(1 - c_{i,\text{out}}^b / c_{i,\text{in}}^b\right) + \text{UF} \cdot c_{i,\text{out}}^b / c_{i,\text{in}}^b \qquad (4.6)$$

The first term in Eq. (4.6) describes the diffusive mass transfer caused by a concentration gradient and the second term the convective mass transfer due to a pressure gradient from blood to dialysate, thereby resulting in ultrafiltration. To explain the significance of both, assume that at $Q_{\text{in}}^b = 200\text{ml/min}$ and $\text{UF} = 8\text{ml/min}$ the urea clearance is $\text{Cl}_{\text{urea}} = 180\text{ml/min}$. At an urea inlet concentration of $c_{\text{urea,in}}^b = 1g/l$, a mass flow of $\dot{M}_{\text{urea}} = Cl_{\text{urea}} \cdot c_{\text{urea,in}}^b = 0.18\,\text{g/min}$ urea is removed from blood into dialysate and a mass flow of $0.02g/\text{min}$ urea will return to the patient with a concentration of $c_{\text{urea,out}}^b = 0.1042g/l$. The diffusive part of the urea clearance is thus about 179.16 ml/min and the convective part 0.84 ml/min.

If, as in hemofiltration, permeable components are only transported by convection, the concentration remains unchanged on blood and filtrate side ($c_{i,\text{in}}^b = c_{i,\text{out}}^b$) and convective clearance results from ultrafiltration.

$$\text{Cl}_{i,\text{konv}} = \dot{M}_{i,\text{konv}} / c_{i,\text{ein}}^b = \text{UF} \qquad (4.7)$$

To achieve the same urea clearance for hemofiltration as in the dialysis example above, ultrafiltration should be as high as 180 ml/min, which in practice is hardly to achieve (see Sect. 4.1.2).

If the convective part of the clearance is neglected ($\text{UF} = 0\,\text{ml/min}$), the inflowing and outflowing volume flows are equal ($Q_{\text{in}}^b = Q_{\text{out}}^b = Q^b$ and $Q_{\text{in}}^d = Q_{\text{out}}^d = Q^d$), and for the diffusive clearance of a component i follows:

$$\text{Cl}_{i,\text{diff}} = \dot{M}_{i,\text{diff}} / c_{i,\text{in}}^b = Q^b \cdot \left(1 - c_{i,\text{out}}^b / c_{i,\text{in}}^b\right) \qquad (4.8)$$

According to the model of boundary layer-controlled mass transfer in countercurrent processes, the diffusive mass flow of a component i through the membrane can be calculated from Eq. (3.55):

$$\dot{M}_{i,\text{diff}} = \dot{m}_{i,\text{diff}} \cdot A^m = K_{0\text{im}} \cdot A^m \cdot \Delta c_{\text{im}} \qquad (4.9)$$

and therefore, diffusive clearance of component I becomes directly proportional to the overall mass transfer coefficient $K_{0\text{im}}$:

$$\text{Cl}_{i,\text{diff}} = \left(K_{0\text{im}} \cdot A^m \cdot \Delta c_{\text{im}}\right) / c_{i,\text{in}}^b \qquad (4.10)$$

The extent to which manufacturers and operators may influence clearance, and thus the efficiency of treatment, consequently depends on the importance of $K_{0\mathrm{im}}$. In Chap. 3.4 (Eq. 3.56), it was deduced that the overall mass transfer coefficient $K_{0\mathrm{im}}$ is related to the blood-side mass transfer coefficient β_{im}^b, the dialysate-side mass transfer coefficient β_{im}^d and the membrane diffusive permeability P_i^m as follows:

$$1/K_{0\mathrm{im}} = 1/\beta_{\mathrm{im}}^b + 1/P_i^m + 1/\beta_{\mathrm{im}}^d \qquad (4.11)$$

The equation for the mean mass transfer coefficient in laminar flow hollow fibre membranes (see Eq. 3.35), adapted to the application with blood, is:

$$\beta_{\mathrm{im}}^b = 1.614 \cdot \left(D_{i,j}^2 \cdot w^B / (d_i \cdot L_{\mathrm{eff}}) \right)^{1/3} \qquad (4.12)$$

If both sides of this equation are expanded with the quotient $d_i / D_{i,j}$, the following relationship is obtained:

$$\beta_{\mathrm{im}}^b \cdot d_i / D_{i,j} = 1.614 \cdot \left(D_{i,j}^2 \cdot w^b \cdot d_i^3 / (d_i \cdot L_{\mathrm{eff}} \cdot D_{i,j}^3) \right)^{1/3} \qquad (4.13)$$

Conversion and expansion of the bracket expression in numerator and denominator with the kinematic viscosity of the liquid phase v_j^b result in:

$$(\beta_{\mathrm{im}}^b \cdot d_i / D_{i,j}) = 1.614 \cdot \left[(w^b \cdot d_i / v_j^b) \cdot (v_j^b / D_{i,j}) \cdot (d_i / L_{\mathrm{eff}}) \right]^{1/3} \qquad (4.14)$$

The variables in Eq. (4.14), which are summarised in parentheses, describe dimensionless numbers. The expression in parentheses on the left-hand side of the equation is the mean Sherwood number $\mathrm{Sh}_{\mathrm{im}}^b$, the 1st bracket expression in the square bracket stands for the Reynolds number Re^b, the 2nd expression in the square bracket is the Schmid number Sc_j^b, and finally there remains a dimensionless ratio of the geometric parameters inner diameter d_i and effective length of the hollow fibre membrane L_{eff}. The outcome of this result in the following equations is:

$$\mathrm{Sh}_{\mathrm{im}}^b = (\beta_{\mathrm{im}}^b \cdot d_i / D_{i,j}) = 1.614 \cdot \left[\mathrm{Re}^b \cdot \mathrm{Sc}_j^b \cdot (d_i / L_{\mathrm{eff}}) \right]^{1/3} \qquad (4.15)$$

$$\mathrm{Re}^b = w^b \cdot d_i / v_j^b \qquad (4.16)$$

$$\mathrm{Sc}_j^b = v_j^b / D_{i,j} \qquad (4.17)$$

Due to the non-circular cross section of the flow channel on the dialysate side (outside the hollow fibre bundle), the hydraulic diameter in the flow cross section of the dialysate space is derived as an equivalent diameter, analogous to the pore diameter of the membrane skin (see pore model, Eq. 3.1):

$$d_h^d = 4 \cdot A_q / U = \left(D^2 - N \cdot d_{\mathrm{out}}^2 \right) / (D + N \cdot d_{\mathrm{out}}) \qquad (4.18)$$

$$\dot{M}_i = Q_{\text{in}}^b \cdot \left(c_{i,\text{in}}^b - c_{i,\text{out}}^b\right) + \text{UF} \cdot c_{i,\text{out}}^b \qquad (4.5)$$

If this mass flow $\dot{M}_i$ is related to the entry concentration of a regarded component i into the dialyser $c_{i,\text{in}}^b$, the result is the *clearance of component i*, a quantity that describes the proportion of the inlet blood flow Q_{in}^b that is completely freed from this component:

$$\text{CL}_i = \dot{M}_i / c_{i,\text{in}}^b = Q_{\text{in}}^b \cdot \left(1 - c_{i,\text{out}}^b / c_{i,\text{in}}^b\right) + \text{UF} \cdot c_{i,\text{out}}^b / c_{i,\text{in}}^b \qquad (4.6)$$

The first term in Eq. (4.6) describes the diffusive mass transfer caused by a concentration gradient and the second term the convective mass transfer due to a pressure gradient from blood to dialysate, thereby resulting in ultrafiltration. To explain the significance of both, assume that at $Q_{\text{in}}^b = 200\text{ml/min}$ and $\text{UF} = 8\text{ml/min}$ the urea clearance is $\text{Cl}_{\text{urea}} = 180\text{ml/min}$. At an urea inlet concentration of $c_{\text{urea,in}}^b = 1g/l$, a mass flow of $\dot{M}_{\text{urea}} = Cl_{\text{urea}} \cdot c_{\text{urea,in}}^b = 0.18\,\text{g/min}$ urea is removed from blood into dialysate and a mass flow of $0.02g/\text{min}$ urea will return to the patient with a concentration of $c_{\text{urea,out}}^b = 0.1042g/l$. The diffusive part of the urea clearance is thus about 179.16 ml/min and the convective part 0.84 ml/min.

If, as in hemofiltration, permeable components are only transported by convection, the concentration remains unchanged on blood and filtrate side ($c_{i,\text{in}}^b = c_{i,\text{out}}^b$) and convective clearance results from ultrafiltration.

$$\text{Cl}_{i,\text{konv}} = \dot{M}_{i,\text{konv}} / c_{i,\text{ein}}^b = \text{UF} \qquad (4.7)$$

To achieve the same urea clearance for hemofiltration as in the dialysis example above, ultrafiltration should be as high as 180 ml/min, which in practice is hardly to achieve (see Sect. 4.1.2).

If the convective part of the clearance is neglected ($\text{UF} = 0\,\text{ml/min}$), the inflowing and outflowing volume flows are equal ($Q_{\text{in}}^b = Q_{\text{out}}^b = Q^b$ and $Q_{\text{in}}^d = Q_{\text{out}}^d = Q^d$), and for the diffusive clearance of a component i follows:

$$\text{Cl}_{i,\text{diff}} = \dot{M}_{i,\text{diff}} / c_{i,\text{in}}^b = Q^b \cdot \left(1 - c_{i,\text{out}}^b / c_{i,\text{in}}^b\right) \qquad (4.8)$$

According to the model of boundary layer-controlled mass transfer in countercurrent processes, the diffusive mass flow of a component i through the membrane can be calculated from Eq. (3.55):

$$\dot{M}_{i,\text{diff}} = \dot{m}_{i,\text{diff}} \cdot A^m = K_{0\text{im}} \cdot A^m \cdot \Delta c_{\text{im}} \qquad (4.9)$$

and therefore, diffusive clearance of component I becomes directly proportional to the overall mass transfer coefficient $K_{0\text{im}}$:

$$\text{Cl}_{i,\text{diff}} = \left(K_{0\text{im}} \cdot A^m \cdot \Delta c_{\text{im}}\right) / c_{i,\text{in}}^b \qquad (4.10)$$

The extent to which manufacturers and operators may influence clearance, and thus the efficiency of treatment, consequently depends on the importance of K_{0im}. In Chap. 3.4 (Eq. 3.56), it was deduced that the overall mass transfer coefficient K_{0im} is related to the blood-side mass transfer coefficient β_{im}^b, the dialysate-side mass transfer coefficient β_{im}^d and the membrane diffusive permeability P_i^m as follows:

$$1/K_{0im} = 1/\beta_{im}^b + 1/P_i^m + 1/\beta_{im}^d \tag{4.11}$$

The equation for the mean mass transfer coefficient in laminar flow hollow fibre membranes (see Eq. 3.35), adapted to the application with blood, is:

$$\beta_{im}^b = 1.614 \cdot \left(D_{i,j}^2 \cdot w^B/(d_i \cdot L_{eff})\right)^{1/3} \tag{4.12}$$

If both sides of this equation are expanded with the quotient $d_i/D_{i,j}$, the following relationship is obtained:

$$\beta_{im}^b \cdot d_i/D_{i,j} = 1.614 \cdot \left(D_{i,j}^2 \cdot w^b \cdot d_i^3/(d_i \cdot L_{eff} \cdot D_{i,j}^3)\right)^{1/3} \tag{4.13}$$

Conversion and expansion of the bracket expression in numerator and denominator with the kinematic viscosity of the liquid phase v_j^b result in:

$$(\beta_{im}^b \cdot d_i/D_{i,j}) = 1.614 \cdot \left[(w^b \cdot d_i/v_j^b) \cdot (v_j^b/D_{i,j}) \cdot (d_i/L_{eff})\right]^{1/3} \tag{4.14}$$

The variables in Eq. (4.14), which are summarised in parentheses, describe dimensionless numbers. The expression in parentheses on the left-hand side of the equation is the mean Sherwood number Sh_{im}^b, the 1st bracket expression in the square bracket stands for the Reynolds number Re^b, the 2nd expression in the square bracket is the Schmid number Sc_j^b, and finally there remains a dimensionless ratio of the geometric parameters inner diameter d_i and effective length of the hollow fibre membrane L_{eff}. The outcome of this result in the following equations is:

$$Sh_{im}^b = (\beta_{im}^b \cdot d_i/D_{i,j}) = 1.614 \cdot \left[Re^b \cdot Sc_j^b \cdot (d_i/L_{eff})\right]^{1/3} \tag{4.15}$$

$$Re^b = w^b \cdot d_i/v_j^b \tag{4.16}$$

$$Sc_j^b = v_j^b/D_{i,j} \tag{4.17}$$

Due to the non-circular cross section of the flow channel on the dialysate side (outside the hollow fibre bundle), the hydraulic diameter in the flow cross section of the dialysate space is derived as an equivalent diameter, analogous to the pore diameter of the membrane skin (see pore model, Eq. 3.1):

$$d_h^d = 4 \cdot A_q/U = \left(D^2 - N \cdot d_{out}^2\right)/(D + N \cdot d_{out}) \tag{4.18}$$

The numerator describes the difference of the empty housing cross section (with the housing diameter D) and the total cross section of all hollow fibre membranes in the module housing (with the number of hollow fibres N and the outer diameter of a hollow fibre d_{out}), the denominator the total wetted circumference. For the dimensionless parameters on the dialysate side, this hydraulic diameter of the dialysate flow channel has to be used in the dialysate Reynolds and Sherwood numbers:

$$Re^d = w^d \cdot d_h^d / v_j^d \tag{4.19}$$

$$Sc_j^d = v_j^d / D_{i,j} \tag{4.20}$$

Analogous to the blood side, a "general from" of a Sherwood equation can be given for the dialysate side, in which proportionality factor "a" and exponent "b" are to be determined for each module and bundle design experimentally:

$$Sh_{im}^d = \beta_{im}^d \cdot d_h^d / D_{i,j} = a \cdot \left(Re^d\right)^b \cdot \left[Sc_j^d \cdot (d_h^d / L_{eff})\right]^{1/3} \tag{4.21}$$

From clearance measurements for molecules usually given in dialyser data sheets, good correlations between experiment and theory have been obtained for Polyflux dialysers from Baxter, if the following correlation factors for the dialysate-side Sherwood number are used:

$$Sh_{im}^d = 0.74 \cdot \left(Re^d\right)^{1.2} \cdot \left[Sc_j^d \cdot (d_h^d / L_{eff})\right]^{1/3} \tag{4.22}$$

For both Sherwood equations, on the blood and dialysate side, it is true that the correlation coefficients determined from experiments are initially values without physical significance. However, further experiments show that for the dialysate-side mass transfer, dependencies of these coefficients on changes in the flow channel by increasing the packing density or the degree of undulation (intensity of the crimping of the hollow fibres), or by special measures on the dialysate side in the housing are possible. When comparing the two Sherwood approaches, it is noticeable that the exponent of the dialysate-side Reynolds number ($b = 1.2$) is clearly higher than that of the blood-side Reynolds number ($b = 1/3$). Although the tangential flows on both sides of the membrane are laminar, the higher exponent of the dialysate-side Reynolds number indicates that the flow around the crossing and undulated hollow fibres causes an improvement of the dialysate-side mass transfer coefficient.

Using the mass balance of a component in steady-state operation, the overall mass transfer coefficient for a component K_{0im} can be calculated by combining Eqs. (4.8), (4.9) and (4.10). From the Sherwood relations (4.15) and (4.22), the mass transfer coefficients β_{im}^b and β_{im}^d are obtained. The diffusive permeability P_i^m may then be calculated from Eq. (4.11). The reciprocal values of these coefficients are, according to Eq. (3.57), the mass transfer resistances R_i^b, R_i^m and R_i^d.

For the discussion of some influencing variables on the diffusive mass transfer through membranes in dialysers, Fig. 4.4 compares results for two dialysers from

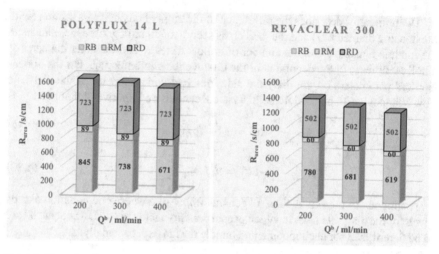

Fig. 4.4 Mass transfer resistances for urea in two different dialysers (Revaclear 300 and Polyflux 14L) as a function of inlet blood flow calculated with values from data sheets of both dialysers. Printed with kind permission of Baxter Inc

Baxter, which were calculated with the equations derived here for the purely diffusive mass transfer of urea at different blood flows and a constant dialysate flow of 500 ml/min.

The Lowflux dialyser Polyflux 14L (see data sheet: HCDE2491_2 © 2009.03. Gambro Lundia AB) contains an asymmetric hollow fibre membrane with an internal diameter of 215 μm and a wall thickness of 50 μm.

In the Highflux dialyser Revaclear 300 (see data sheet HCDE15649_2 © 2013.03. Gambro Lundia AB), the inner diameter is 190 μm and the wall thickness 35 μm. The membrane area is 1.4 m^2 for both dialysers.

The results for the Polyflux 14L show that the dialysate transport resistances R_{urea}^d give constant values of 723 s/cm, which should be the case by maintaining the dialysate flow of 500 ml/min. When blood flow is increased from 200 to 400 ml/min, blood transport resistance R_{urea}^b decreases from 845 to 671 s/cm, and the membrane transport resistances R_{urea}^m stay constant at 89 s/cm. Therefore, the total transport resistances decrease from 1,657 to 1,483 s/cm, which means that the overall mass transfer coefficient of urea increases by about 11% if the blood flow is increased from 200 to 400 ml/min.

The transport resistances for urea in the Revaclear 300 are all lower than those in the Polyflux 14L. The membrane resistance is smaller because of the smaller wall thickness and because of the better diffusive properties in the larger pores of the Highflux dialyser. The blood resistance is reduced compared to the Polyflux 14L because of the smaller inner diameter and the improved mass transfer coefficient in the hollow fibres due to the higher blood velocity. The dialysate resistance is lower than that of the Polyflux 14L due to improved flow around the fibres through an optimised bundle structure.

Fig. 4.5 Influences of blood, dialysate flow and ultrafiltration on clearance of urea and vitamin B12 (values taken from the data sheet of Revaclear 300). Printed with kind permission of Baxter Inc

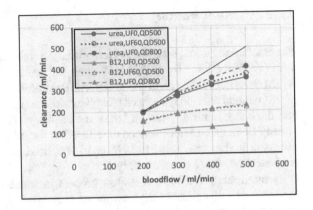

As a result of these measures, the total transport resistance in the Revaclear 300 is also significantly lower and thus the mass transfer coefficient and the clearance are significantly higher than in the Polyflux 14L under the same operating conditions.

The potential for improving diffusive clearance for urea therefore lies in optimising the blood- and dialysate-side flow guidance and in measures to reduce membrane resistance. If the same considerations are made for a larger molecule, such as vitamin B12, the membrane resistance dominates over the boundary layer resistances, and this of course to a considerable extent with the Lowflux membrane of the Polyflux 14L.

A further potential for increasing the removal rates is an additional convective transport by higher ultrafiltration rates. A plot of the clearance values of urea and vitamin B12 from the data sheet of the Revaclear 300 for different operating conditions (see Fig. 4.5) shows that the increase in clearance of small molecules, such as urea, by additional, convective transport (UF = 60 ml/min) is low (2–4%), whereas this can be increased by approximately 10% for larger molecules, such as vitamin B12. Similarly, high changes in clearance values result in vitamin B12 if the dialysate flow is increased from 500 to 800 ml/min when UF = 0 ml/min. The best choice with respect to performance would be to do both high UF and high dialysate flow. However, costs for the treatment by preparation and disposal of a significantly higher dialysate volume and for about 10–15 l sterile infusion fluid will be much higher.

When using *Lowflux* membranes, diffusive transport dominates for small molecules. With *Highflux* membranes, convective transport is particularly advantageous for larger molecules, although this effect is partly reduced by the occurrence of back filtration within a module. The mathematical description of the superimposed transport processes when using *Highflux* membranes is not possible within the scope of this introduction; therefore, reference is made to the corresponding literature (Wuepper (1997), Raff (2003), Eloot (2004)).

4.1.2 Hemofiltration

The first "International Meeting on Hemofiltration" took place in Berlin in December 1981 (Schäfer (1982). Synthetic membranes had been developed for this application for the first time at the end of the 1960s. At the conference in Berlin, the advantages of this method over dialysis, such as higher circulatory and blood pressure stability and the better removal rates for larger molecules, were reported. There were also already findings on arteriovenous ultrafiltration for dehydrating patients with fluid overload. In the years between 1975 and 1985, hemofiltration was developed as an equivalent treatment to haemodialysis.

Hemofiltration (HF) is a crossflow process in which fluid is removed from the blood by setting a suitable transmembrane pressure. The separation characteristics of the membranes developed for this purpose correspond to those used for *Highflux* dialysis. This means that molecules with molecular masses of about 10–20 kD can also be removed in HF. In this process, ultrafiltration is regulated by infusion of a sterile substitution solution into the arterial (predilution) or venous (post-dilution) blood tubing. If the infusion volume flow is adjusted in a way that about 2 l net in 4 h is withdrawn (i.e. about 8.3 ml/min), at an ultrafiltration of 150 ml/min the infusion flow rate must therefore be 141.7 ml/min. At these high ultrafiltration rates, compared to dialysis, proteins are transported convectively to the membrane surface and a concentration boundary layer will be formed on top of the membrane (at the membrane wall). Accordingly, the dependencies of the target variables UF can be described with the model of boundary layer-controlled mass transfer in a crossflow process (see Chap. 3.2).

As an example, results from the data sheet of the "Prismaflex System Hemofilter Set" from Baxter, which is used for acute renal failure in intensive care units with a process called Continuous Veno-Venous Hemofiltration (CVVH) will be discussed here (see Fig. 4.6).

The course of ultrafiltration above the transmembrane pressure determined by in vitro tests with adjusted bovine blood (hct = 32%, protein concentration: 60 g/l)

Fig. 4.6 Ultrafiltration (UF) as a function of transmembrane pressure (TMP) for different inlet blood flows (Qb) in the hemofilter Prismaflex M100 (ref. Data sheet Prismaflex M100 Set; 306100279_1_2009.09. Gambro Lundia AB). Printed with kind permission of Baxter Inc

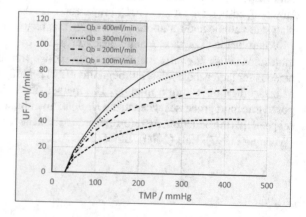

shows a comparable course to the experiments by Neggaz (2007) with aqueous albumin solutions (see Chap. 3.2, Fig. 3.5). In the region of a fully developed boundary layer, a further increase in TMP no longer leads to an increase in UF. Comparing the courses at different blood flows, this region of boundary layer-controlled mass transfer shifts to higher transmembrane pressures at higher blood flows. In the investigated TMP range for Fig. 4.6, a plateau may clearly be recognised only for $Q_m^b = 100$ ml/min.

As described in Chap. 3.2, the course of the curves can be divided into three regions. In the example at transmembrane pressures between 30 and 50 mmHg, it can be assumed that no concentration boundary layer is formed. The transport resistance in this TMP range is the hydraulic membrane resistance R_h^M. Therefore, the relationship between UF and TMP$_m$ in this region can be described with the pore model (see Chap. 3.1).

$$UF = UFC \cdot (TMP_m - \pi_{onc.})\qquad(4.23)$$

The osmotic pressure difference between blood and filtrate side may be assumed as the oncotic pressure of the rejected proteins.

According to the model, the region, where UF is independent of TMP$_m$, will be reached when the concentration boundary layer of the impermeable components is completely developed and dominates the transport resistance. For this region, the following applies in modified form ($\dot{V}^P$ = UF index "f" = index "b") of Eq. (3.37):

$$
\begin{aligned}
UF &= A^m \cdot \beta_{im}^b \cdot \ln(c_{iw}^b/c_{ibm}^b) \\
&= A^m \cdot 1.614 \cdot \left(D_{i,j}^2 \cdot w_m^b/(d_i \cdot L_{eff})\right)^{1/3} \cdot \ln(c_{iw}^b/c_{ibm}^b)\qquad(4.24)
\end{aligned}
$$

The influences of operating, physical and geometric variables are derived from the equation for the blood-side mass transfer coefficient recommended by Leveque (1928) for laminar flow β_{im}^b (Eq. (3.35)). For the treatment of a patient with the same hemofilter (same geometric data), the influencing factors are reduced to the mean blood flow velocity:

$$w_m^b = 4 \cdot Q_m^b/(N \cdot \pi \cdot d_i^2) = 2 \cdot (Q_{in}^b + Q_{out}^b)/\left(N \cdot \pi \cdot d_i^2\right)\qquad(4.25)$$

The UF in this TMP-independent region can therefore only be increased by a higher mean blood flow Q_m^b (resp. inlet blood flow Q_{in}^b). In the model, this means that the blood-side mass transfer coefficient β_{im}^b increases, due to higher shear rates causing boundary layer thickness δ_{im}^b to decrease.

The ultrafiltration curves for the M100 set (Fig. 4.6) show that an increase in ultrafiltration with blood flow is already clearly pronounced in the transition area between membrane-controlled and boundary layer-controlled mass transfer. It therefore makes sense to set the blood flow as high as possible already in this region, described by the combined equation according to Neggaz (2007) (see Eq. (3.43)).

Fig. 4.7 Permeate flow density (ultrafiltration rate) as a function of protein bulk concentration in hemofiltration at different blood flow velocities. *Source* Goehl et al. (1982), Printed with kind permission of Karger

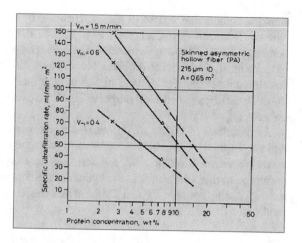

Measurements by Goehl et al. (1982), presented in Fig. 4.7 in a semi-logarithmic diagram as a function of $UF/A^m = f(c_{ibm}^b)$ according to the following equation

$$UF/A^m = \beta_{im}^b \cdot \ln(c_{iw}^b/c_{ibm}^b) = \beta_{im}^b \cdot \ln c_{iw}^b - \beta_{im}^b \cdot \ln c_{ibm}^b, \qquad (4.26)$$

show a linear course with the slope $-\beta_{im}^b$ and the x-axis intercept c_{iw}^b at $UF/A^m = 0$. The influence of blood flow velocity $w_m^b = v_M$ is confirmed since the ultrafiltration flow density UF/A^m increases with increasing velocity at the same bulk concentration.

Since UF in a hemofilter is usually adjusted to a value of 20–30% of the inlet blood flow to avoid a high haematocrit in the venous line back to the patient, the clearance values for low-molecular weight components in hemofiltration are significantly lower than the values achieved in dialysis even at high blood flows (at $Q_{in}^b = 500$ml/min, $UF = Cl_{conv} = 100$ to 150 ml/min). Components with higher molar mass, as β_2-microglobulin, will be cleared in hemofiltration in the order of magnitude as in *Highflux* dialysis. Comparing the treatment costs, sterile substitution fluid as used in HF in high volumes (30–40 l) is significantly higher than those for the non-sterile dialysate. Since *Highflux* dialysis offers the advantages of high clearances for the whole range of permeable uremic toxins at moderate costs, hemofiltration is only rarely used to treat patients with end-stage renal disease (ESRD). However, CVVH is becoming increasingly important, especially for the treatment of acute organ failure in ICUs, also due to the SARS-CoV-2 pandemic (see Fig. 1.1).

4.2 Plasma Donation and Plasma Exchange

Plasma donation with membranes has been used since the beginning of the 1980s to donate one's own plasma for planned surgery with rather high blood loss thereafter, but also to use it as a source for plasma components (albumin, immunoglobulins, coagulation factors, etc.). Since the blood cells are returned to the donor and only plasma is collected, there is no need for a waiting time of about 6 weeks that must be observed for whole blood donation (Heal (1983)).

Plasma exchange, to reduce pathologically increased concentrations of proteins in the plasma, such as LDL, cholesterol, immune complexes, or toxins bound to proteins, has been used meanwhile in a high number of applications for different diseases (Malchevsky (2004), Szcepiorkowski (2010), Winters (2013)).

The "MONET" system has been developed by Fresenius for the treatment of patients with hypercholesterolaemia and will be explained here as one example. The principle of the cascade filtration used in this process is shown in Fig. 4.8. Patient's blood is pumped through a plasma filter, and by applying a TMP, plasma is filtered while the blood cells are rejected. Without the additional MONET filter, the removed plasma would have to be substituted with sterile human plasma when returning the blood to the patient.

For this process, the MONET filter has been developed to reject 67% of the "bad" low-density lipoprotein (LDL), 52% of the triglycerides and 38% of the cholesterol and excrete them in a waste bag, while the "valuable, small" proteins such as albumin, the immunoglobulins IgG and IgM and the "good" high-density lipoprotein (HDL) may largely pass through the MONET membrane and are returned to the patient via the venous blood line. The ultrafiltration rate in the MONET filter is significantly influenced by a boundary layer of rejected proteins and can be described with the same equations as derived for hemofiltration in Sect. 4.1.2.

To estimate the achievable filtration rate of the plasma filter, and thus the achievable LDL removal rate, models are used in which the formation of a blood cell layer is considered. According to the particle layer model by Rippberger (1993) (see Chap. 3.3), ultrafiltration for plasma filters is calculated according to Eq. (3.46):

$$\text{UF} = \dot{V}^p = A^m \cdot \text{TMP}_m / \left(\eta^l \cdot \left(R_h^m + R^{pl} \right) \right) \tag{4.27}$$

After conversion, from this equation results for the particle layer resistance:

$$R^{pl} = A^m \cdot \text{TMP}_m / \left(\text{UF} \cdot \eta^l \right) - R_h^m \tag{4.28}$$

The value for the hydraulic membrane resistance R_h^m is calculated according to the pore model, from Eq. (3.13):

$$R_h^m = 1 / \left(\eta^l \cdot L_P \right) \tag{4.29}$$

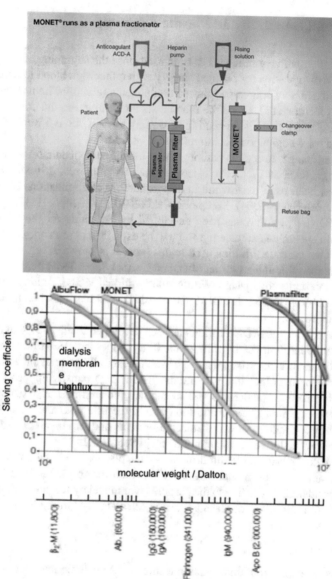

Fig. 4.8 Cascade filtration in plasmapheresis for the treatment of LDL patients (*Source* Brochure Therapeutic Apheresis Fresenius 733 995 1/1 D (1 PUR 12.06) and sieving coefficients of selected Fresenius membranes (Source Therapeutic Apheresis Fresenius 733 995 1/1 D (1 PUR12.06). Printed with kind permission of Fresenius

Fig. 4.9 Plasma filtration rate of the Prismaflex plasma filter 2000 as a function of transmembrane pressure and blood flow, measured with bovine blood at 37 °C, HKt = 32%, protein conc.: 60 g/l (*Source* data sheet: Prismaflex 2000 set: HCEN4307_6 © 2010.07. Gambro Lundia AB). Printed with kind permission of Baxter Inc

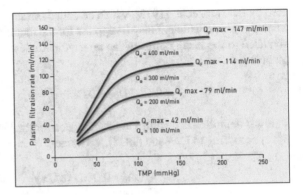

Assuming no cells block the pores, the membrane resistance R_h^m adjusts to a constant value after plasma contact (see Chap. 3.1, Eloot experiments). The cell layer resistance R^{pl} increases with the mean transmembrane pressure TMP_m and the associated changes in the particle (cell) layer parameters (thickness, porosity, tortuosity; see Eq. (3.45)) until a value for ultrafiltration is reached at which the cell layer thickness does not increase furthermore. In this state, the maximum cell layer thickness on the membrane surface is reached for the adjusted operating parameters blood flow and *TMP*.

Using experimental results from the data sheet of the Prismaflex Plasmafilter 2000 from Baxter with a membrane area of 0.35 m² (see Fig. 4.9), the resistance values may be estimated as follows. Assuming at the highest inlet blood flow of 400 ml/min the shear rate in the hollow fibres to be high enough to avoid at low transmembrane pressures a particle layer formation, hydraulic permeability of the membrane with plasma at 37 °C ($\eta^L(37°C) = 0.0022\,Pa*s$) results from the slope of this line to $L_P = 7.16 \cdot 10^{-10}\,$m/(Pa · s) and the hydraulic membrane resistance to $R_h^m = 6.35 \cdot 10^{11}\,$m^{-1}.

The slopes of the curves for blood flows between 100 and 300 ml/min in the region of low transmembrane pressures indicate that cell layer deposits started already to develop on the membrane surface. If the particle layer resistance in the linear range for an inlet blood flow of 100 ml/min is determined from Eq. (4.28), using the hydraulic membrane resistance R_h^m calculated above, the result for the particle layer resistance in this region becomes $R^{pl} = 6.91 \cdot 10^{11}\,$m^{-1} and is thus in the same order of magnitude as the hydraulic membrane resistance. After formation of the maximum cell layer thickness, this value increases to $1.75 \cdot 10^{12}\,$m^{-1}.

By such calculations, correlations between UF and operating conditions (TMP, Q^b, shear rate) can be derived. Rippberger (1993) also investigated influences of particle size and porosity of the formed particle layers. However, these approaches are not suitable for optimising the module/membrane geometries (length, diameter, number of hollow fibre membranes).

Therefore, based on the model of boundary layer-controlled mass transfer in crossflow operation, Zydney (1982) proposed to replace the diffusion coefficient of dissolved components according to Stokes–Einstein (see Eq. (3.39)) by a definition used by Eckstein et al. (1972), which he called the shear-enhanced diffusion coefficient D_γ. This coefficient may be applied for the mobility of particles in liquids and is related to the mean radius of the rejected particles a_i and the wall shear rate γ_w according to:

$$D_\gamma = 0.025 \cdot a_i^2 \cdot \gamma_w \tag{4.30}$$

If this definition is included to the approach for the local mass transfer coefficient $\beta_i^b(x)$ according to Leveque (1928), we obtain:

$$\beta_i^b(x) = 0.046 \cdot \left(a_i^4/x\right)^{1/3} \cdot \gamma_W \tag{4.31}$$

With the wall shear rate in a laminar flow profile

$$\gamma_w = 8 \cdot w_m^b/d_i, \tag{4.32}$$

and the mean blood velocity

$$w_m^b = 4 \cdot Q_m^b/N \cdot \pi \cdot d_i^2, \tag{4.33}$$

ultrafiltration according to the model of shear-enhanced diffusion results after integration over the length of the hollow fibres in:

$$\mathrm{UF} = 2.24 \cdot a_i^{4/3} \cdot Q_m^b \cdot L_{\mathrm{eff}}^{2/3}/d_i^2 \cdot \ln(c_{\mathrm{iw}}^{\mathrm{bv}}/c_{\mathrm{ibm}}^{\mathrm{bv}}) \tag{4.34}$$

Under the assumptions that in applications with blood, mainly the erythrocytes, with a radius of about $a_i = 4\mu m$, make up the cell layer on the wall, the mean bulk concentration c_{ibm}^b corresponds to the mean blood haematocrit (hct). The value for the wall concentration is proposed to be $c_{\mathrm{iw}}^{\mathrm{bv}} = 0.95$, corresponding to hct $=$ 0.95. If these values are included in Eq. (4.34), dependence of ultrafiltration on operational, geometric and patient-specific erythrocyte parameters can be expressed by the following equations:

$$\mathrm{UF} = K(\mathrm{hct}) \cdot Q_m^B \cdot L_{\mathrm{eff}}^{2/3}/d_i^2 \tag{4.35}$$

$$K(\mathrm{hct}) = 2.24 \cdot a_i^{4/3} \cdot \ln(0.95/\mathrm{hct}) \tag{4.36}$$

According to this model, the ultrafiltration achievable in the plateau during plasmapheresis increases with the mean blood flow, with increasing effective length, decreasing inner diameter of the hollow fibre membranes and at low hct.

In the Zydney model (1982), reference is made to a limit load in plasmapheresis above which the erythrocytes are damaged by excessive shear forces and lead to haemolysis. This is mentioned in the data sheets for plasma filters by the recommendation of manufacturers that maximum transmembrane pressures for defined blood flows should not be exceeded for given hollow fibre and module geometries. These values are based on the perception that the haematocrit at the exit from the plasma filter should not exceed a value of 65%, because at this high value in the blood that is returned to the patient, rupture of the erythrocytes and thus haemolysis can occur due to shear forces.

In the example of the Baxter plasma filter (see Fig. 4.9), if the achievable filtration rate in the plateau is 79 ml/min at an inlet blood flow of 200 ml/min, the hct increases to about 66% at an inlet value of 40% in the exiting blood with a flow of $200-79 = 121$ ml/min. The TMP_{max} given in the data sheet for these operating conditions is 171 mmHg. As can be seen from Fig. 4.9, at this TMP and an inlet blood flow of 200 ml/min, the UF plateau is reached, so further increasing of the TMP is also not reasonable from efficiency reasons. It can be concluded that the geometric parameters of the hollow fibre membranes L_{eff} and d_{in} in this plasma filter are optimised with respect to the UF plateaus achievable at the blood flow-dependent TMP_{max} values but also for preventing haemolysis.

Comparing the two models, Rippberger's extended particle layer model can be used to calculate the UF for the entire TMP range, with a surface layer becoming thicker and more concentrated with TMP. The slope $dUF/dTMP$ is decreasing with TMP until the UF plateau is reached. Higher blood flows increase the shear rate and decrease the layer thickness, resulting in higher UF plateaus.

Although the Zydney model may only be used for the calculation of the UF—plateau values, it can be applied to optimise the geometric data of a module (length and inner diameter of the hollow fibre membranes) for a safe and efficient treatment of patients.

4.3 Artificial Liver: MARS and PROMETHEUS

The functions of the human liver are so complex that the artificial liver processes available today can only take over a very small fraction of them. As with the artificial kidney, these are essentially the detoxification functions. In contrast to the kidney, in the liver mainly hydrophobic toxins, which are protein-bound and dissolved in the blood plasma, are disposed of by decoupling from the carrier protein with subsequent hydrophilisation of the toxins via bile and intestine. One such toxin is indirect, unconjugated bilirubin, a degradation product of the red blood pigment, haemoglobin. The bilirubin cycle in Fig. 4.10 shows that bilirubin is released in the spleen and introduced into the bloodstream as indirect bilirubin by binding to albumin. The complex of albumin and bilirubin is separated in the liver. Albumin is returned to the bloodstream with the now free binding site, and bilirubin is coupled to glucuronic acid and thus converted to direct, water-soluble bilirubin, which can be excreted via bile and intestine or via the kidneys.

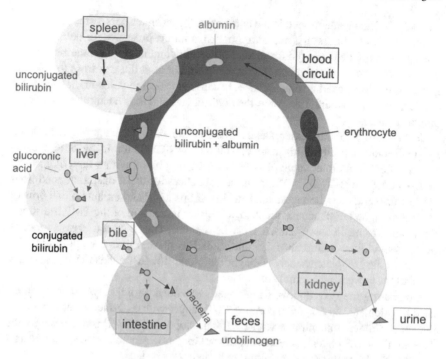

Fig. 4.10 Bilirubin circuit. Printed with kind permission of Martin Mißfelder

To implement this pathway in an artificial liver, it is necessary to enable both physico-chemical steps, cleavage of toxins from the carrier protein and excretion of bilirubin in the membrane process.

MARS.

Stange (1993) reported the first successes of a process to remove protein-bound drugs and toxins by dialysis with albumin in a closed dialysate circuit in which toxin-loaded albumin is continuously recycled. From this work, the "molecular adsorbent recirculating system (MARS)" (see Fig. 4.11) was developed in cooperation between the Rostock-based company Teraklin, which produced the MARS monitor, and the company Gambro (now Baxter), with the task of developing a membrane optimised for this application.

The separation of the toxin from the albumin in the "MARSFlux dialyser" is achieved by partial dissociation of the albumin–toxin complex. The free toxin molecule migrates diffusively through the membrane skin and meets a solution with human albumin on the dialysate side, in which the toxin is taken up by binding to free binding sites. The formed albumin–toxin complex in the "albumin circuit" is separated via activated carbon adsorption and ion exchange, in which the toxin is either adsorbed on activated carbon by van der Waals forces or attached to suitable ion exchange resins by ionic interactions in exchange with counter ions. Regenerated

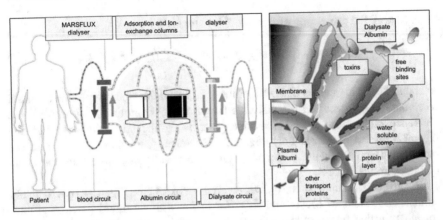

Fig. 4.11 Schematic of an *Artificial Liver* according to the "molecular adsorption recirculation system (MARS)" (left) and transport model for protein-bound toxins across the membrane of the MARSFlux dialyser (right). Printed with kind permission of Baxter Inc

albumin flows back to the dialysate inlet of the "MARSFlux dialyser" where it can again binds a free toxin molecule. If both the patient's liver and kidney have lost their function, an integrated, conventional dialyser can take over the function of the "Artificial Kidney".

This "albumin dialysis" with the MARS is a counter-current process, and thus, the equations derived in Chap. 3.2 and Sect. 4.1.1 apply to describe the transmembrane mass transport, respectively the clearance. These equations however assume freely moving molecules (urea, creatinine, electrolytes, etc.) in plasma. To show differences in the removal rates of such uremic toxins with protein-bound toxins, Meyer et al. (2004) determined clearance values for urea, creatinine, and phenol red (PR) for the dialyser Optiflux F200NR from Fresenius with an "artificial plasma", *initially without albumin,* against a standard dialysate. The clearance value for phenol red with a molar mass of 354 g/mol was lower than the clearances of urea (molar mass: 60 g/mol) and creatinine (molar mass: 113 g/mol), but with a value of 158 ml/min at $Q^b = 205$ ml/min, $Q^d = 289$ ml/min, compared to 182 ml/min for urea and 183 ml/min for creatinine in the order of magnitude expected for the clearly higher molecular PR.

After albumin had been added to the "artificial plasma", the clearance values for urea and creatinine remained almost unchanged, while the clearance of PR (14 ml/min) was reduced by about 90% compared to that without albumin in the plasma. This difference arises because PR is bound to albumin with an equilibrium constant of:

$$K_{A-PR} = c_{PR,b}/\left[c_{PR,f} \cdot \left(c_A - c_{PR,b}\right)\right] = 2.8 \cdot 10^4 \mathrm{M}^{-1} \qquad (4.37)$$

and only the free portion $c_{PR,f}$ can permeate the membrane. With the values for K_{A-PR} and the concentrations for albumin c_A and phenol red c_{PR} used by Meyer et al. (2004), the proportion of free PR f is calculated from:

$$f = c_{PR,f}/(c_{PR,f} + c_{PR,b}) = c_{PR,f}/c_{PR} = 1/[1 + (c_A - c_{PR}) \cdot K_{A-PR}] \quad (4.38)$$

to about 0.06. This means that 94% of the total PR in the albumin plasma is bound to albumin. By increasing the dialysate flow from 290 ml/min to 740 ml/min, the clearance values for urea and creatinine could be significantly increased, while the clearance of PR improved only slightly. Consequently, transport of bound toxins cannot be described by the model of boundary layer-controlled mass transport in the counter-current process.

In vitro experiments by Raff (2006) with the MARS (see Fig. 4.12) qualitatively confirm the results of Meyer and show for toxins with varying degrees of affinity for albumin that transport through the membrane is essentially dependent on the binding forces between protein and toxin, expressed by the binding constant K_A. In the experiments with plasma enriched by albumin-bound toxins (ABTs), the concentrations recorded over time are related to the measured values 30 min after the start of the experiment, because after this time steady-state conditions (fluxes, pressures, temperature) had set in. The concentrations in the plasma pool at blood and dialysate

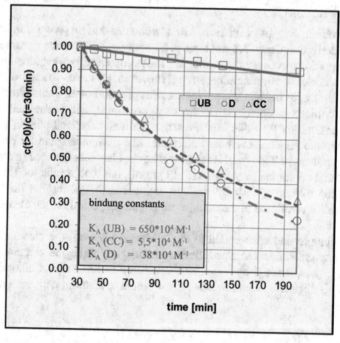

Fig. 4.12 Removal rates for unconjugated bilirubin (UB), chenodeoxycholic acid (CC) and diazepam (D) in the MARS (*Source* presentation by Raff, Euromembrane 2006; see Raff et al. 2006). Printed with kind permission of Baxter Inc

flows of 250 ml/min each reach at 150 min (corresponding to 120 min treatment time) about 92% for unconjugated bilirubin (UB), about 35% for diazepam (D) and about 41% for chenodeoxycholic acid (CC) of the values reached after 30 min. In Meyer's experiment, the PR concentrations were about 65 and 55% of the starting concentration after 120 min of experiment, at dialysate flows of 300 and 750 ml/min, respectively.

Compared to the diffusive transport of free toxins in the artificial kidney, the mass transfer coefficients in the boundary layers of blood β_{im}^b and dialysate β_{im}^d in the MARS may only be considered for the less strongly bound toxins. Decisive for the transport of liver toxins with high affinity to albumin (large binding constants), such as bilirubin, in the MARS system is the diffusive membrane permeability P_i^M.

Comparative measurements with different membrane materials by Reimann et al. (1995) show that a suitable distribution of hydrophilic and hydrophobic domains in the polymer gives the best results compared to purely hydrophilic or purely hydrophobic polymers. *UF membranes made of hydrophobic polymers* (e.g.*polyamide*) do remove hydrophobic toxins, but only until the binding sites for the toxins on the membrane are saturated. The binding between toxin and polymer is so strong that further transport through the pores is not possible. *Dialysis membranes made of hydrophilic polymers (e.g. cuprophane)* show no removal of hydrophobic toxins. The repulsive, hydrophilic–hydrophobic interactions are apparently large enough to reject even dissociated, free toxin molecules. The best removal rates could be achieved with a membrane produced from a *polymer alloy containing hydrophobic polyether sulfone (PES) and hydrophilic polyvinylpyrrolidone (PVP)*. The desired ratio of hydrophilic–hydrophobic domains at the membrane surface, which is optimal for the interactions between the ABT complex and the membrane surface, can be realised by suitable settings of the precipitation parameters when spinning the hollow fibre membranes.

The asymmetric cross-sectional structure of the membrane wall was chosen in a way that dialysate-side albumin can penetrate through the large-pored outer skin of the membrane into the area of the vacuoles and already "pick up" toxins there (see Fig. 4.11). Then, fewer free toxins will enter the dialysate stream, and the mass transfer resistance on the dialysate side R_i^d will be negligible for such membrane structures. The diffusive membrane resistance P_i^m which is inversely proportional to the diffusion path (length of the membrane pore) can also be reduced with this special membrane structure.

The interpretation of the importance of the membrane structure for the MARS procedure is difficult because the specific binding forces between toxin and carrier protein on the one hand and between toxin and membrane polymer on the other hand are decisive for diffusive transport (free diffusion, surface diffusion, etc.) and this cannot be optimised equally for the very different toxins. Therefore, the treatment parameters in acute liver failure with the MARS will be oriented towards the components that are most difficult to remove (e.g. UB).

PROMETHEUS.

The "PROMETHEUS" system offered by Fresenius to support liver function follows a different concept (see Fig. 4.13). The membrane in the "Albuflow filter" used in it has a significantly higher MWCO than that in the MARS filter (being a Highflux dialyser membrane) and is designed in such a way that albumin–toxin complexes and other proteins of this size can pass through the membrane (see Fig. 4.8, Sect. 4.3). In this crossflow process, transport through the membrane is convective, analogous to plasmapheresis. The clearance of the "albumin-bound toxins (ABTs)" is determined here by the UF. To estimate the achievable ultrafiltration rate of the Albuflow, the models of the transport equations for the particle-layer-controlled mass transport may be used (see Chap. 3.3 and Sect. 4.3). It should however be considered that the ABTs in the filtered plasma are separated by adsorption to a neutral resin (Albuflow AF01) and to an anion exchange resin (Albuflow AF02) and toxin-free patient plasma albumin flows into the filtrate space of the Albuflow via back filtration. The pressure curves on the blood and dialysate sides in the Albuflow will therefore interlace, as in Highflux dialysis, so in comparison with the MARS counter-current process, the membrane surface in the PROMETHEUS crossflow process must be optimised with respect to forward and back filtration.

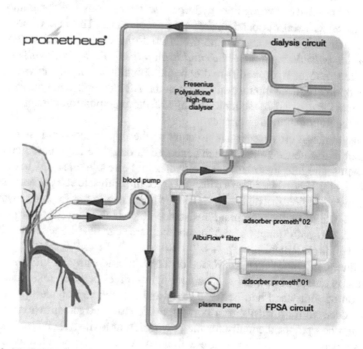

Fig. 4.13 PROMETHEUS therapy system by Fresenius to support liver function (*Source* Data sheet Fresenius: Prometheus: 732 042 1/2 GB (1 GUT 12.07)). Printed with kind permission of Fresenius

For the achievable toxin clearance, in addition to the balance of ultrafiltration and back filtration in the "Albuflow", the removal kinetics of the toxins in the adsorbers (the proportion of albumin freed from the toxin) must also be recorded.

If necessary, a dialyser can be inserted in the venous blood tubing back to the patient, which takes over the removal of uremic toxins and thus enables a combination of *Artificial Liver* and *Artificial Kidney*, as in the MARS.

Comparing the two systems, in the PROMETHEUS procedure the albumin–toxin complex is removed from the patient's blood and regenerated on the filtrate side in a serial connection of adsorber and ion exchanger and toxin-free-albumin is returned to the patient via back filtration in the Albuflow filter.

In the MARS, the albumin–toxin complex in the patient's blood is decoupled in the hollow fibres of the MARSFlux dialyser and only the toxin permeates through the membrane to the dialysate side of the MARSFlux dialyser. There, it is taken up by "foreign" human albumin and transported to adsorber and ion exchanger columns in the "albumin circuit", where it is bound. Regenerated albumin is returned to the dialysate side of the MARSFlux dialyser to take up further toxin molecules.

It may be assumed that the removal rate is higher with the PROMETHEUS method than with the MARS, especially for toxins with high binding constants. However, this also means that a proportion of the patient's blood plasma is in contact with adsorbents and ion exchanger resins, which can lead to losses of valuable substances (proteins, hormones, etc.) through non-specific accumulation. According to Nevens and Laleman (2012), both types of treatment are well tolerated by patients with "acute liver failure (ALF)", especially to bridge the time until liver transplantation.

4.4 Artificial Lung

The lungs are the organ supplying blood with oxygen (O_2) from the air we breathe in and disposing of carbon dioxide (CO_2) via the air we breathe out. The blood is pumped from the right ventricle of the heart through the pulmonary arteries into the pulmonary capillaries, where the absorption of O_2 and the desorption of CO_2 take place.

In contrast to the process for the *Artificial Kidney* and the *Artificial Liver*, in which substances are exchanged between two liquid phases, in the Artificial Lung the components pass a gas–liquid phase boundary.

As shown in Chap. 3.4.2 to avoid discontinuity in the concentration course over a gas–liquid interface, the partial pressure is used as a concentration measure. Then, instead of the Henry coefficient, which is inversely proportional to the concentration, Bunsen absorption coefficient α_i is used as a proportionality factor between the volume concentrations of component i in the liquid phase related to the partial pressure.

$$c_i^{lv} = \alpha_i\,(T) \cdot p_i^l \tag{4.39}$$

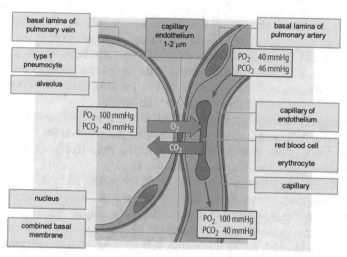

Fig. 4.14 Exchange of oxygen (O_2) and carbon dioxide (CO_2) in the lung (*Source* Faller (1999)). Printed with kind permission of Thieme

Figure 4.14 shows the mean partial pressures for O_2 and CO_2 in the gas phase (alveolus) and in the liquid phase (blood inlet and blood outlet) of a human lung. The oxygen partial pressure in the blood inlet to the lung, in the pulmonary artery, up to the mass transfer in the area of the "combined basal membrane" is on average 40 mmHg, which, with a Bunsen coefficient for O_2 in blood at 37 °C of $\alpha_{O_2}^b$ (37°C) $= 0.028$ ml $O_2/$(l blood $*$ mmHg), corresponds to a concentration of 1.12 ml $O_2/$l blood. At the exit of the blood from the lung, in the pulmonary vein, the concentration of the purely physically dissolved oxygen at an O_2 partial pressure of $p_{O_2}^b = 100$ mmHg is 2.80 ml $O_2/$l blood.

This increase in the partial pressure of O_2 in the blood plasma from 40 to 100 mmHg by purely physical absorption would mean that 8.4 ml $O_2/$min would pass from the gas phase into the blood at a blood flow of 5 l/min.

However, since the oxygen consumption of a healthy adult human at rest is about 250–300 ml $O_2/$min, significantly more oxygen must be transferred to blood than corresponds to the physically dissolved values. This is achieved by the reaction of dissolved oxygen in plasma with haemoglobin (Hb), which is predominantly contained in the erythrocytes, to form $Hb(O_2)_n$. The proportion of $Hb(O_2)_n$ increases with the partial pressure of O_2 and reaches an O_2 saturation of almost 100% at about 100 mmHg (see Fig. 4.15). In this process, 1.34 ml $O_2/$g Hb ("Hüfner's number") is bound, which means that at a normal value of the Hb concentration in the blood of 150 g Hb/l, the oxygen bound to haemoglobin is 201 ml $O_2/$l blood. At a partial pressure for oxygen of 40 mmHg, in the pulmonary artery, the Hb saturation is only about 75% and the concentration of bound oxygen will be 150 ml $O_2/$l blood. Thus, to achieve full Hb-O_2 saturation of 201 ml $O_2/$l blood at a blood flow of 5 l/min, an O_2 volume flow of 255 ml $O_2/$min must pass the basal membranes of all alveoli.

4.4 Artificial Lung

Fig. 4.15 Saturation curve
of haemoglobin with oxygen
according to Yoshida (1993)
(bl solid line, Eq. (4.46)).
Linearised saturation
intervals (blue, red and
yellow lines) with partial
pressure mean values
(dashed lines) for each region

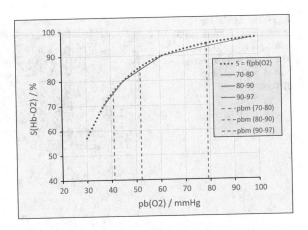

Oxygen in the blood is consumed by oxidation of carbon-containing compounds in the body cells, and among other molecules, CO_2 is formed as a reaction product. The partial pressure of CO_2 in the cells thus becomes higher than that in the blood, which leads to a diffusive transport of CO_2 through the membranes of the tissue cells into the intercellular space and into the blood. This increases the CO_2 partial pressure in the blood from the mean arterial value of about 40 mmHg to a mean venous value of about 46 mmHg. Venous blood is pumped through the left heart muscle to the right ventricle, from where it is transported through the pulmonary artery to the lungs. Using the partial pressures of CO_2 according to Fig. 4.14 and a Bunsen coefficient of $\propto_{CO_2}^{b}$ (37°C) $= 0.645$ ml CO_2/(l blood*mmHg), the CO_2 concentrations in the blood at the lung inlet are 29.7 ml CO_2/l blood and at the lung outlet 25.8 ml CO_2/l blood. At a blood flow of 5 l/min, 19.5 ml CO_2/min of physically dissolved CO_2 would therefore be transferred to the expiratory air. However, by hydrolysis reactions in plasma and red cells CO_2 is also found as bicarbonate (HCO_3^-) and carbonate (CO_3^{2-}) ions, all together resulting in CO_2 concentration of 480 ml CO_2/l in arterial and 530 ml CO_2/l in venous blood (Schmidt 2010).

Thus, if as in a heart surgery the lung function should be completely taken over by an *Artificial Lung*, at a blood flow of 5 l/min about 255 ml O_2/min should be taken up from an adequate gas mixture and about the same flow of 250 ml CO_2/min should be released from blood into the gas phase.

O_2 - Mass transfer without Hb

In contrast to the human lung, in an *Artificial Lung*, flows through gas and blood compartments and the gas phase composition can be individually adjusted. In the extracorporeal life support (ECLS) process, the composition of the gas phase is a mixture of compressed air and oxygen.

For early membrane oxygenators which were operating in a counter-current mode and where blood flows inside the hollow fibres and gas outside, the mass transfer coefficient can be determined again from the Leveque equation as described in Chap. 3.4.2. With the mean blood flow velocity in the hollow fibres

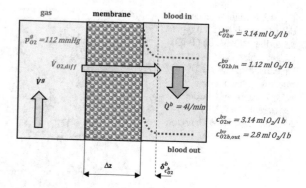

Fig. 4.16 Model for the transport of O_2 from gas to liquid phase (blood) in a membrane oxygenator with a porous, hydrophobic membrane

$$w^b_m = 4 \cdot \dot{Q}^b / \left(N \cdot \pi \cdot d_i^2 \right), \tag{4.40}$$

from Eqs. (3.70) and (3.71), the O_2 mass transfer coefficient in the blood film results in

$$\beta^b_{O_2 m} = 1.75 \cdot \left(D^2_{O_2,b} \cdot \dot{Q}^b / (N \cdot d_i^3 \cdot L_{\text{eff}}) \right)^{1/3}, \tag{4.41}$$

and the transmembrane O_2 volume flow becomes

$$\dot{V}_{O_2,\text{diff}} = 1.75 \cdot \left(D^2_{O_2,b} \cdot \dot{Q}^b / (N \cdot d_i^3 \cdot L_{\text{eff}}) \right)^{1/3} \cdot A^m \cdot \Delta c^{bv}_{O_2,m} \tag{4.42}$$

In Fig. 4.16, values for concentrations and volume flows are given as they might occur in a counter-current oxygenator. The partial pressure of O_2 in the gas phase $p^g_{O_2}$ is assumed to be 112 mmHg. The bulk O_2 concentrations at the inlet and outlet of the oxygenator are the values calculated from the respective partial pressures of O_2 in the blood, 40 mmHg at the inlet and 100 mmHg at the outlet. With a blood flow of 4 l/min, the transmembrane O_2 volume flow results from the O_2 balance to: $\dot{V}_{O_2,\text{diff}} = 4\,l/min*(2.8-1.12)$ ml $O_2/l = 6.72$ ml O_2/min.

Under these conditions, 6.72 ml O_2/min would pass the membrane and will be absorbed in the liquid phase. In this purely physical process (without Hb), the partial pressure of O_2 at the outlet would be adjusted to the desired value of 100 mmHg. With a partial pressure of O_2 in the gas phase of 112 mmHg, the equilibrium concentration in the blood at the phase boundary interface is calculated to be 3.14 ml O_2/l. The mean logarithmic concentration difference in counter-current oxygenation according to Eq. (3.67) results in:

$\Delta c^{bv}_{O_2 m} = (2.8-1.12)$ ml $O_2/l/ln\ ((3.14-1.12)/(3.14-2.8)) = 0.94$ ml O_2/l.

With an assumed membrane surface area of 4.4 m^2, which is the size of a dialyser used by Yoshida (1993) in a counter-current process, the mean mass transfer coefficient, derived from Eq. (3.70)

$$\beta^b_{O_2,m} = \dot{Q}^b \cdot \left(c^{bv}_{O_2 b,\text{out}} - c^{bv}_{O_2 b,\text{in}} \right) / (A^m \cdot \Delta c^{bv}_{O_2,m}), \tag{4.43}$$

is calculated to be

$$\beta_{O_2 m}^b = = 2.71 * 10^{-5} \text{ m/s} = 0.163 \text{ cm/min}$$

O_2 - Mass transfer including reaction with Hb

According to Yoshida (1993) and Wickramasinghe (2005), the model of purely physical absorption of O_2 can be extended to include the chemical reaction between haemoglobin and oxygen by multiplying the average mass transfer coefficient by an *enhancement factor E*, to get:

$$\beta_{O_2,\text{eff}}^B = E \cdot \beta_{O_2,m}^B \tag{4.44}$$

This *enhancement factor* defined by Yoshida shows the following dependencies of partial pressure of O_2 in the gas phase ($p_{O_2}^g$ in mmHg), haematocrit ht and Hb-O_2 saturation S:

$$E = \left(714/p_{O_2}^g\right)^{1/3} \cdot \left[1 + 11.8((1-S) \cdot \text{ht})^{0.8} - 8.9 \cdot ((1-S) \cdot \text{ht})\right] \tag{4.45}$$

The course of the O_2-Hb binding curve can be described by the following equation (Hill 1910):

$$S = \left(p_{O_2}^b/p_{O_2,50}^b\right)^{2.7} / \left(1 + \left(p_{O_2}^b/p_{O_2,50}^b\right)^{2.7}\right) \tag{4.46}$$

The O_2 partial pressure at which 50% of the binding sites of the Hb are saturated results from Eq. (4.46) in $p_{O_2,50}^b = 27$ mmHg. Due to the nonlinear, sigmoidal course of $S = f(p_{O_2}^b)$, it is not reasonable to calculate the O_2 transport across the membrane with an averaged concentration gradient over the entire length of an oxygenator. Yoshida (1993, example 2, Table 1) proposed to subdivide the saturation curve in the range between the O_2 partial pressures 40 and 100 mmHg according to Fig. 4.15 into 3 linear regions. Within these regions, a differential O_2 volume flow $\Delta \dot{V}_{O_2}$ with a mean partial pressure $p_{O_2 m}^b$ can be calculated for a differential membrane area ΔA^m from a modified Eq. 3.70 extended by the enhancement factor:

$$\Delta \dot{V}_{O_2} = \Delta A^m \cdot E \cdot \beta_{O_2,m}^b \cdot (c_{O_2,w}^{bv} - c_{O_2,b}^{bv})$$
$$= \Delta A^m \cdot E \cdot \beta_{O_2,m}^b \cdot \alpha_{O_2} (p_{O_2}^g - p_{O_2 m}^b) \tag{4.47}$$

Based on the maximum O_2 saturation of $c_{hb-O_2,\text{max}}^b = 210$ ml O_2/l at a haemoglobin concentration of 140 mg/l, the local O_2 volume flow for the saturation of the Hb from 70 to 80% at a blood flow of 4 l/min results in:

$$\Delta \dot{V}_{O_2,\text{diff},70-80} = \dot{Q}^b \cdot \Delta S/100 \cdot c_{hb-O_2,\text{max}}^b = 84 \text{ ml } O_2/\text{min} \tag{4.48}$$

If Eq. (4.47) is solved for the differential membrane area, it follows:

$$\Delta A^m = \Delta \dot{V}_{O_2,\text{diff}}/(E \cdot \beta_{O_2,m}^b \cdot \alpha_{O_2} (p_{O_2}^g - p_{O_2 m}^b)) \tag{4.49}$$

The mean O_2 partial pressure $p^b_{O_2m}$ in the blood between 70 and 80% O_2-Hb saturation is $p^b_{O_2m}(S_m = 75\%) = 41$ mmHg. The O_2 partial pressure at the phase boundary between blood and gas phase after reaching phase equilibrium corresponds to the O_2 partial pressure in the gas, in Yoshida's example $p^g_{O_2} = 714$ mmHg.

From Eq. (4.45), with a haematocrit of 40% and a mean saturation of $S = 0.75$, the enhancement factor is $E = 1.97$.

With the physical data

$\propto_{O_2} = 2.82*10^{-5}$ ml O_2/(ml*mmHg),

$D^b_{O_2} = (2.13 - 0.92 \cdot ht/100) \cdot 10^{-5} cm^2/s = 1.76 \cdot 10^{-9} m^2/s$,

and the geometrical data for the assumed oxygenator with a membrane surface area of 4.4 m^2 (as mentioned above, these data are chosen to compare calculations here with that described by Yoshida (1993, example 2, Table 1)

$$N = 50000, \; d_i = 200 \cdot 10^{-6} m, \; L_{eff} = 0.14m,$$

at a blood flow of $\dot{Q}^b = 4$ l/min, the mass transfer coefficient results from Eq. (4.49) in

$\beta^b_{O_2m} = 0.163$ cm/ min $= 4.53 \cdot 10^{-5}$ m/s.

The differential membrane area required for an O_2 volume flow of 84 ml O_2/min in the saturation interval between 70 and 80% is then:

$$\Delta A^m_{70-80} = 1.38 m^2$$

Similarly, the required membrane areas for the O_2 volume flow of again 84 ml O_2/min in the saturation interval between 80 and 90% are calculated to give $\Delta A^m_{80-90} = 1.63$ m^2, and for an O_2 volume flow of 58 ml O_2/min in the saturation interval between 90 and 97%, it results in $\Delta A^m_{90-97} = 1.39$ m^2. Consequently, for a total O_2 volume flow of 226 ml O_2/min, a membrane area of 4.4 m^2 is required in a counter-current oxygenator. This relatively large membrane area of an oxygenator operating in a counter-current process causes high costs and means a higher risk in terms of biocompatibility.

The comparison of counter-current and crossflow heat exchangers had shown that significantly higher heat transfer coefficients will be reached if liquids flow perpendicular to each other. Therefore, oxygenators have been developed in which blood flows outside the hollow fibre membranes interlinked as mats perpendicular to the gas flow inside the hollow fibres (see Fig. 4.17).

Experiments by Wickramasinghe (2005) with such oxygenators from Cobe (e.g. Optima XP, corresponding to module D in Fig. 4.17) show in double logarithmic order of $Sh^b_{O_2m}/(Sc^b)^{1/3}$ as a function of Re^b a good fit of the experimental results (see Fig. 4.18) for the following Sherwood relationship.

$$Sh^b_{O_2m} = \beta^b_{O_2m} \cdot d_E/D^b_{O_2} = 0.8 \cdot \left(Re^b\right)^{0.59} \cdot (Sc^b)^{1/3} \qquad (4.50)$$

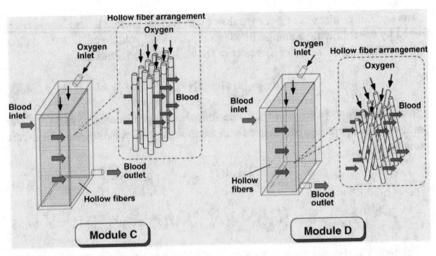

Fig. 4.17 Hollow fibre arrangements and blood flow direction in membrane oxygenators *Source* Nagase et al. (2005). Printed with kind permission of Elsevier

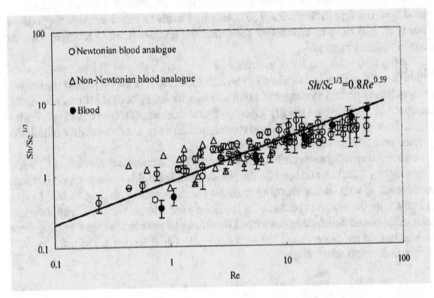

Fig. 4.18 Correlation of experimental results for the mass transfer of oxygen in hollow fibre oxygenators (*Source* Wickramasinghe (2005)). Printed with kind permission of J. Wiley & Sons

$$\mathrm{Re}^b = w^b \cdot d_E / \nu^b \tag{4.51}$$

$$\mathrm{Sc}^b = \nu^b / D_{\mathrm{O}_2}^b \tag{4.52}$$

The equivalent diameter d_E for the flow cross section outside the hollow fibres is defined by the following equation in which ε is the void volume outside the hollow fibre bundle and d_{out} the outer diameter of the hollow fibre membranes:

$$d_E = [\varepsilon/(1 - \varepsilon)] \cdot d_{\text{out}} \tag{4.53}$$

From these Eqs. (4.50) to (4.52), the following dependencies of mass transfer coefficient on operating parameters and membrane and module geometries are obtained:

$$\beta_{O_2m}^b = 0.8 \cdot \left(\text{Re}^b\right)^{0.59} \cdot (\text{Sc}^b)^{1/3} \cdot D_{O_2}^b/d_E \tag{4.54}$$

$$\beta_{O_2m}^b = 0.8 \cdot \left(\left(w^b\right)^{0.59} \cdot \left(D_{O_2}^b\right)^{0.67}\right) / \left(d_E^{0.41} \cdot \left(v^b\right)^{0.26}\right) \tag{4.55}$$

The mean blood-side mass transfer coefficient can therefore be increased by increasing blood velocity w^b, by improving the diffusivity of O_2 ($D_{O_2}^b$) within the blood boundary layer and by reducing the equivalent diameter d_E, respectively the porosity ε of the blood flow channel. The higher exponent of Reynolds number or blood flow velocity (0.59 versus 0.33) compared to the Leveque solution for counter-current oxygenators indicates that the crossflow process is more efficient than the counter-current process.

For the Cobe Optima XP oxygenator ($A^m = 1.9$ m², $d_E = 480$ mm), the required total O_2 volume flow is calculated to be 226 ml O_2/min at an O_2 partial pressure in the gas phase of 280 mmHg and at an average blood velocity of 4 cm/s using the approach of Yoshida (see above). The blood-side mass transfer coefficient is 3.7 times higher than in the counter-current oxygenator, allowing a much smaller membrane surface area for the same performance.

By adjusting diffusion coefficient and viscosity, the equations according to Wickramasinghe (2005) can be applied for non-Newtonian fluids such as blood or blood analogues, as well as for Newtonian fluids (such as water; see Fig. 4.18). Consequently, with the general approach for the Sherwood number (Eq. 4.56), applications with other media and further developed modules and membranes may also be tested and optimised by determining the best correlation coefficients a and b for the new design and/or the other media.

$$\text{Sh}_{\text{im}} = \beta_{\text{im}} \cdot d_E/D_{\text{ij}} = \text{a} \cdot (\text{Re})^{\text{b}} \cdot (\text{Sc})^{1/3} \tag{4.56}$$

With this understanding, developments in the design of the membrane bundle structure in oxygenators led to the realisation of the required exchange volume flows for O_2 to replace lung function with a relatively small membrane surface area (2-3m²) compared to that of all alveoli in the lungs (approximately 90m²).

The derived relationships may also be used for devices that require lower exchange rates than oxygenation during cardiac surgery, such as for patients with limited lung function due to COVID-19, chronic obstructive pulmonary disease (COPD), interstitial lung disease (ILD) and others.

CO_2 - Mass transfer including chemical reactions

As shown above if lung function should completely be replaced by an artificial lung, a volume flow of about 250 ml CO_2/min should be released from blood into the gas phase. The calculation of CO_2 mass transfer is analogous to that for O_2. The driving force is a concentration gradient from the blood to the gas phase (in the negative z-direction).

According to Katoh (1978), transport of dissolved CO_2 in plasma is supported by the simultaneous diffusion of bicarbonate ions (HCO_3^-) which are converted into CO_2 and H_2O by dehydration. The good fit of the theoretical CO_2 rates with the values experimentally determined on oxygenator prototypes with hydrophobic membranes confirms this theory.

A summary of Sherwood correlations for different kinds of oxygenator designs is given by Low (2017), but in all references presented there "no information on the development of CO_2 mass transfer correlation has been identified". Therefore, the group developed numerical approaches to describe the mass transfer for both O_2 and CO_2 in non-Newtonian blood streams for square and staggered arrangements of hollow fibre membranes used in cross- (transvers) and in counter-current (parallel) processes, considering physically dissolved and chemically bound fractions of O_2 and CO_2. By correlating theoretical with experimental results, the following Sherwood equations for the staggered hollow fibre arrangement in a transvers flow process have been derived for O_2 and CO_2.

$$Sh_{O_2m}^b = 0.1311\varepsilon^{-0.666}\left(Re^b\right)^{0.3433\varepsilon^{-0.034}}\left(Sc_{O_2}^b\right)^{1/3} \tag{4.57}$$

$$Sh_{CO_2m}^b = 0.5216\varepsilon^{-0.505}\left(Re^b\right)^{0.2547\varepsilon^{-0.130}}\left(Sc_{CO_2}^b\right)^{1/3} \tag{4.58}$$

From these Sherwood correlations, mass transfer coefficients $k = \beta$ for O_2 and CO_2 have been calculated for hollow fibres with an outer diameter of 380 μm and a blood flow of 5 l/min. The results are shown in Fig. 4.19 as a function of the void fraction ε of the blood flow cross section outside the hollow fibres. For all modifications, k decreases with increasing ε. For CO_2 (blue lines), higher mass transfer coefficients were reached than for O_2 (red lines). Crossflow (Transv.) processes show higher k-values for both gases than counter-current (Paral.) flow processes.

These results show, that due to the higher mass transfer coefficients of CO_2 compared to O_2, the design of oxygenators optimised on the mass transfer of O_2 meets the requirements for sufficient CO_2 desorption.

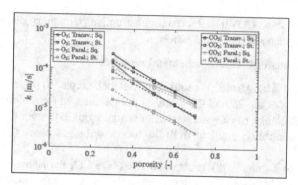

Fig. 4.19 Comparison of mass transfer coefficients of O_2 (red lines) and CO_2 (blue lines) in model oxygenators with square (Sq.) and staggered (St.) fibre arrangement in processes with blood flow directions parallel (Paral.) and perpendicular (Transv.) to the gas flow (*Source* Low (2017)). Printed with kind permission of ASME International

References

Alwall ND, Norvit L, Steins AM (1948) Clinical extracorporeal dialysis of blood with artificial kidney, The Lancet 61–62

Eckstein EC, Bailey DG, Shapiro AH (1972) Self-diffusion of particles in shear flow of suspensions. J Fluid Mech 79:191

Eloot S (2004) Experimental and numerical modelling of dialysis, Ph.D. dissertation, Ghent University

Faller A, Schinke M (1999) Der Körper des Menschen, Thieme Stuttgart

Goehl H, Konstantin O, Gullberg CA (1982) Hemofiltration membranes. Contri Nephr 32:20–30, Karger Basel

Heal JM, Bailey G, Helphingstine C, Thiem PA, Leddy JP, Buchholz DH, Nusbacher J (1983) Non-centrifugal plasma collection using crossflow membrane plasmapheresis. Vox Sang 44(3):143–150

Hill AV (1910) The possible effects of the aggregation of the molecules of haemoglobin on its dissociation curves. J Physiol 40:iv–vii

Katoh S, Yoshida F (1978) Carbon dioxide in a membrane blood oxygenator. Anal of Biomed Eng 6:48–59

Kolff WJ, Higgins C (1954) Dialysis in the treatment of uremia: artificial kidney. J Urol 72(6)

Low, KWQ, van Loon R, Rolland AS, Sienz J (2017) Formulation of generalized mass transfer correlations for blood oxygenator design. J Biomech Eng. Mar 1:139 (3). https://doi.org/10.1115/1.4035535

Malchesky PS et al (2004) Apheresis technologies and clinical applications: the 2002 international apheresis registry. Therapeutic Apheresis and Dialysis 8(2):124–143, Blackwell Publishing, Inc

Meyer TW, Leeper EC, Bartlett DW, Depner TA, Zhao Y, Robertson CR, Hostetter TH (2004) Increasing dialysate flow and dialyzer mass transfer area coefficient to increase the clearance of protein-bound solutes. J Am Soc Nephrol 15:1927–1935

Nagase K, Kohori F, Sakai K (2005) Oxygen transfer performance of membrane oxygenator composed of crossed and parallel hollow fibres. Biochemical Eng J 24:105–113

Nevens F, Laleman W (2012) Artificial liver support devices as treatment option for liver failure. Best Pract Res Clin Gastroenterol 26:17–26

Neggaz Y, Lopez Vargas M, Ould Dris A, Riera F, Alvarez R (2007) A combination of serial resistances and concentration polarization models along the membrane in ultrafiltration of pectin and albumin solutions. Separation Purif Tech 54:18–27

References

Raff M, Welsch M, Göhl H, Hildwein H, Storr M, Wittner B (2003) Advanced modelling of highflux hemodialysis. J Membr Sci 216:1–11

Raff M, Ertl T, Krause B, Storr M, Göhl H (2006) Mass transfer in artificial liver membrane devices. Desalination 199:234–235

Reimann A, Betz S, Raff M (1995) Removal of albumin bound toxins by extended dialysis. Intern J of Art Org 18(8):465

Rippberger S (1993) Berechnungsansatze zur Crossflow-Filtration, Chem 1ng Tech. 65 Nr. 5, S:533–540

Ronco C, Bagshaw SM, Bellomo R, Clark WR, Hussein-Syed F, Kellum JA, Ricci Z, Rimmele T, Reis T, Ostermann M (2021) Extracorporeal blood purification and organ support in the critically ill patient during COVID-19 pandemic: Expert review and recommendation. Blood Purif 50:17–27

Schäfer K, Koch E, Quellhorst E, von Herrath D (1982) Hemofiltration, Contribution Contributions to Nephrology, Karger, ISBN 3-8055-3515 5

Schmidt RF, Lang F, Heckmann M (2010) Physiologie des Menschen. Springer Medizin Verlag Heidelberg. ISBN-13 978-3-642-01650-9

Stange J, Ramelow W, Mitzner S, Schmidt R, Klinkmann H (1993) Dialysis against a recycled albumin solution enables the removal of albumin-bound toxins. Artif Organs 17(9):809–813

Stamatialis DF, Papenburg BJ, Girones M, Saiful S, Bettahalli NM, Schmitmeier S, Wesseling M (2008) Medical applications of membranes: drug delivery, artificial organs and tissue engineering. J Membr Sci 308:1–34

Szczepiorkowski ZM (2010) Guidelines on the use of therapeutic apheresis in clinical practice—evidence-based approach from the apheresis applications committee of the American society for apheresis. J Clin Apheresis 25:83–177

Wickramasinghe SR, Han R, Garcia JD, Specht R (2005) Microporous membrane blood oxygenators. AIChE J 51(2):656–670

Winters JL (2013) Randomized controlled trials in therapeutic apheresis. J Clin Apheresis 28:48–55

Wüpper A, Dellana F, Baldamus CA, Woermann D (1997) Local transport process in high-flux hollow fiber dialyzers from the models. J Membr Sci 131:181–193

Yoshida F (1993) Prediction of oxygen transfer performance of blood oxygenators. Artif Organs Today 2(4):237–252

Zweigart C, Neubauer M, Storr M, Böhler T, Krause B (2010) Progress in the development of membranes for kidney-replacement therapy. In Drioli E, Giorno L (eds) Comprehensive Membrane

Zydney AL, Colton CK (1982) Continuous flow membrane plasmapheresis: theoretical models for flux and hemolysis prediction. Trans Am Soc Art Intern Organs 28:408–412

Chapter 5
Conclusions

Abstract Membrane processes are used in technical and medical applications for the separation of waste from valuable substances. To design these processes for a desired separation efficiency, the required membrane surface area and a suitable mode of operation for the solution/dispersion to be treated must be determined. To derive these relationships for different modes of operation, process-specific models for the transport of a component "i" through a differential membrane area are created and local transport equations are determined. The application of these equations to AO processes was chosen because it addresses a wide range of membrane properties and modes of operation. By integrating the local transport equations over the entire membrane surface area in a module, the desired relationships are obtained. This allows users to optimise the treatment, manufacturers to further develop membrane modules and processes, and students to understand and adapt the approach to other applications.

Biological membranes are indispensable in nature because, as a semi-permeable barrier between compartments, they very efficiently regulate exchange processes in which particles, cells or macromolecules are rejected, and solvents or solutions with valuable and/or waste substances may permeate. The current fields of application of technical membranes in environmental protection (e.g. exhaust air, water purification, etc.), in biotechnology (e.g. concentration of solutes and membrane bioreactors) and in medicine (artificial organs) are correspondingly diverse. The processes used differ in the operating parameters to be selected and the phases and components involved in the mass transfer.

The production of hollow fibre membranes and modules is shortly introduced in Chap. 2 to show the possibilities for changing membrane structure, hollow fibre bundles and dimensions in the module.

Using models on differential membrane elements, functional relationships between desired "target variables" of a process (e.g. filtration and diffusion fluxes), operating variables (e.g. flow velocity and transmembrane pressure), substance data (e.g. viscosity, molar mass, concentration) and geometric data (e.g. inner diameter and wall thickness of hollow fibre) are derived and extended to the total membrane

© The Author(s), under exclusive license to Springer Nature Switzerland AG 2022   69
M. Raff, *Mass Transfer Models in Membrane Processes*,
SpringerBriefs in Bioengineering,
https://doi.org/10.1007/978-3-030-89195-4_5

surface area in a hollow fibre module. These models for different processes are deliberately derived for an unspecified component "i" to indicate that the equations can be applied not only to artificial organs but also to technical membrane processes.

The application of membrane processes to Artificial Organs has been chosen because it allows a wide range of requirements to be shown, both in terms of the separation characteristics of membranes and in terms of interactions between blood and fluids (e.g. dialysate) with the polymers used to make membranes. Therefore, equations developed in Chap. 3 have been adapted in Chap. 4 to the processes in AO, resulting in exchange rates of organ-specific components as functions of parameters mentioned above.

From these, the supervising personnel, trained in medical and equipment technology, can adapt suitable operating parameters to the respective, patient-specific requirements and thus individually optimise the treatment.

Manufacturers of membranes and modules can use the results to improve their products when changes must be considered due to new medical findings or economic constraints.

Interested students, professionals and lecturers with knowledge in fluid mechanics, physical chemistry and process engineering can use the models presented here and adapt them for other applications.

Between the publication of the German version and this book, infections with the novel COVID-19 spread worldwide, causing many elderly patients to become so severely ill that their lung, kidney and liver functions were severely impaired at the same time. Several monitors with very different requirements, multiple monitoring sensors and warning signals were therefore used to support all the diseased organs, making it very difficult for the already heavily burdened staff in the intensive care units to make the right decision in each case (to press the right "buttons").

Efforts are therefore being made to combine the very different criteria for adequate removal rates of organ-specific toxins in one machine and thus to enable intelligent monitors to regulate necessary corrections also regarding interactions (see, e.g., Fuhrmann (2020) with ADVOS). For patients and staff in intensive care units, such "combined artificial organs" will be of immense advantage.

Printed in the United States
by Baker & Taylor Publisher Services